Lecture Notes in N

Series Editors

Fakher Chaari, *National School of Engineers, University of Sfax, Sfax, Tunisia*
Francesco Gherardini, *Dipartimento di Ingegneria "Enzo Ferrari", Università di Modena e Reggio Emilia, Modena, Italy*
Vitalii Ivanov, *Department of Manufacturing Engineering, Machines and Tools, Sumy State University, Sumy, Ukraine*

Subline Series Editor

Francisco Cavas-Martínez, *Departamento de Estructuras, Construcción y Expresión Gráfica Universidad Politécnica de Cartagena, Cartagena, Murcia, Spain*

Editorial Board Members

Francesca di Mare, *Institute of Energy Technology, Ruhr-Universität Bochum, Bochum, Nordrhein-Westfalen, Germany*
Mohamed Haddar, *National School of Engineers of Sfax (ENIS), Sfax, Tunisia*
Young W. Kwon, *Department of Manufacturing Engineering and Aerospace Engineering, Graduate School of Engineering and Applied Science, Monterey, CA, USA*
Justyna Trojanowska, *Poznan University of Technology, Poznan, Poland*
Jinyang Xu, *School of Mechanical Engineering, Shanghai Jiao Tong University, Shanghai, China*

Lecture Notes in Mechanical Engineering (LNME) publishes the latest developments in Mechanical Engineering—quickly, informally and with high quality. Original research reported in proceedings and post-proceedings represents the core of LNME. Volumes published in LNME embrace all aspects, subfields and new challenges of mechanical engineering.

To submit a proposal or request further information, please contact the Springer Editor of your location:

Europe, USA, Africa: Leontina Di Cecco at Leontina.dicecco@springer.com
China: Ella Zhang at ella.zhang@springer.com
India: Priya Vyas at priya.vyas@springer.com
Rest of Asia, Australia, New Zealand: Swati Meherishi at swati.meherishi@springer.com

Topics in the series include:

- Engineering Design
- Machinery and Machine Elements
- Mechanical Structures and Stress Analysis
- Automotive Engineering
- Engine Technology
- Aerospace Technology and Astronautics
- Nanotechnology and Microengineering
- Control, Robotics, Mechatronics
- MEMS
- Theoretical and Applied Mechanics
- Dynamical Systems, Control
- Fluid Mechanics
- Engineering Thermodynamics, Heat and Mass Transfer
- Manufacturing
- Precision Engineering, Instrumentation, Measurement
- Materials Engineering
- Tribology and Surface Technology

Indexed by SCOPUS and EI Compendex.

All books published in the series are submitted for consideration in Web of Science.

To submit a proposal for a monograph, please check our Springer Tracts in Mechanical Engineering at https://link.springer.com/bookseries/11693

Athanasios Chasalevris · Carsten Proppe
Editors

Advances in Active Bearings in Rotating Machinery

Proceedings of The 1st Workshop on Active Bearings in Rotating Machinery (ABROM 2022), June 29–30, 2022, Athens, Greece

Springer

Editors
Athanasios Chasalevris
School of Mechanical Engineering
National Technical University of Athens
Athens, Greece

Carsten Proppe
Institute of Engineering Mechanics
Karlsruhe Institute of Technology
Karlsruhe, Germany

ISSN 2195-4356 ISSN 2195-4364 (electronic)
Lecture Notes in Mechanical Engineering
ISBN 978-3-031-32393-5 ISBN 978-3-031-32394-2 (eBook)
https://doi.org/10.1007/978-3-031-32394-2

© The Editor(s) (if applicable) and The Author(s), under exclusive license to Springer Nature Switzerland AG 2023

This work is subject to copyright. All rights are solely and exclusively licensed by the Publisher, whether the whole or part of the material is concerned, specifically the rights of translation, reprinting, reuse of illustrations, recitation, broadcasting, reproduction on microfilms or in any other physical way, and transmission or information storage and retrieval, electronic adaptation, computer software, or by similar or dissimilar methodology now known or hereafter developed.

The use of general descriptive names, registered names, trademarks, service marks, etc. in this publication does not imply, even in the absence of a specific statement, that such names are exempt from the relevant protective laws and regulations and therefore free for general use.

The publisher, the authors, and the editors are safe to assume that the advice and information in this book are believed to be true and accurate at the date of publication. Neither the publisher nor the authors or the editors give a warranty, expressed or implied, with respect to the material contained herein or for any errors or omissions that may have been made. The publisher remains neutral with regard to jurisdictional claims in published maps and institutional affiliations.

This Springer imprint is published by the registered company Springer Nature Switzerland AG
The registered company address is: Gewerbestrasse 11, 6330 Cham, Switzerland

Preface for 1st ABROM Proceedings

Active bearings are an emerging technology in rotating machines, and they are considered as a subset of active components for controlling machine dynamics. This area of engineering combines advanced physics and technical solutions, mostly achieved by mechatronic elements, smart materials, and control. Rotating machines and rotating components exist in a wide range of applications, from large low-speed marine propulsion systems, to small high-speed spindles for manufacturing processes; in between, one may consider numerous applications like large and small steam/gas/hydro/wind turbines for power generation, aircraft engines, electric drives and generators, turbochargers for automotive/marine applications in internal combustion engines, turbopumps for space propulsion and oil/gas extraction, microturbines, and others.

The active control of the dynamics of rotating machines and rotating systems is not new. Shortly after electronics were embedded in machines, active magnetic bearings (AMBs) appeared as a promising technology to substitute traditional solutions for lubrication and support of moving/rotating components. However, the need for dedicated hardware and software for the operation of AMBs together with cost and reliability parameters still retains the traditional bearing elements such as sliding bearings and ball bearings sufficiently competitive.

Active bearings follow the growth of active components in machines of Industry 4.0 era. There are numerous contributions from pioneers to this objective, and from these, one might conclude that the recent developments in the field gather the principles of oil/water/gas lubrication, the recent advances in automatic control, machine learning, parameter estimation and health monitoring, tribology, and material science, aiming to establish a standard technology that guides future machines in ultra-high speeds, higher efficiency, and downsizing, including the feature of adaptivity in operational conditions. Therefore, active bearings are perceived as a key component in cyberphysical systems in rotating machines.

In 2020, another project in the field of active bearings was initiated as a research synergy between the Institute of Engineering Mechanics of the Karlsruhe Institute of Technology and the Laboratory for Dynamics and Acoustics of the National Technical University of Athens. The group linkage program focused on nonlinear dynamics of rotors with active bearings and was funded by the *Alexander von Humboldt Foundation* in Germany; one of the requirements of AvH Foundation was the organization of a Workshop for the dissemination of the results. The two group heads decided to invite pioneers in the objective, and it is an honor for the organizers that reference scientists were present at the event, solely in physical presence; this is how the 1st ABROM Workshop was born. The Workshop included 14 paper presentations and 28 participants from 9 countries. Twelve papers were submitted for publication and are included in this book of proceedings after a standard peer-review process.

The editors would like to acknowledge *Alexander von Humboldt Foundation* for the decisive contribution to the realization of the 1st ABROM Workshop.

The editors dedicate this book of proceedings to the memory of *Prof. Wolfgang Seemann* (Karlsruhe Institute of Technology—Institute of Engineering Mechanics) who passed away few months before the 1st ABROM Workshop takes place.

Athanasios Chasalevris
Carsten Proppe

Contents

Rotordynamic Modal Testing via Journal Bearing Lubricant
Film – An Experimental Validation 1
   Ilmar Ferreira Santos and Mads Nowak Frigaard

Problems of Development of Tribotronics: Control and Machine Learning
Techniques 12
   *Leonid Savin, Denis Shutin, Yuri Kazakov, Alexey Kornaev,
Roman Polyakov, and Shengbo Li*

A Presentation of Control Theory Applied to the Design of Controllable
Segmented Gas Foil Bearings 27
   Janus Walentin Jensen and Ilmar Ferreira Santos

Tilting-Pad Journal Bearing with Active Pads: A Way of Attenuating
Rotor Lateral Vibrations 44
   Heitor A. P. da Silva and Rodrigo Nicoletti

Studying the Effect of Viscous Friction Minimization in Actively
Lubricated Journal Hybrid Bearings 55
   Denis Shutin, Leonid Savin, and Yuri Kazakov

On the Feedback Control of a Rotor System with Active Flexible Bearings 65
   Alexander Bitner and Carsten Proppe

A Novel Monolithic Shape-Morphing Bearing for Real-Time NVH Control 82
   *Christos Kalligeros, Georgios Chantoumakos, Efstratios Tsolakis,
Panagiotis Spiridakos, and Vasilios Spitas*

Application of Squeeze Film Dampers 111
   Edoardo Gheller, Steven Chatterton, Andrea Vania, and Paolo Pennacchi

Investigation of Active Configuration in Gas Foil Bearings for Stable Ultra
High-Speed Operation 134
   Ioannis Gavalas, Anastasios Papadopoulos, and Athanasios Chasalevris

Gyroid Lattice Structures for Tilting Pad Journal Bearings 150
   *Ludovico Dassi, Steven Chatterton, Paolo Parenti, Andrea Vania,
and Paolo Pennacchi*

Behavior of an Active Magnetic Bearing as a Stern Tube Bearing: A First Approach via Simulations 162
  Vasileios-Menelaos Koufopanos and Pantelis G. Nikolakopoulos

On the Stability Margins of Parametrically Excited Rotating Shafts on Gas Foil Bearings: Linear and Nonlinear Approach 188
  Emmanouil Dimou, Fadi Dohnal, and Athanasios Chasalevris

Author Index 209

Rotordynamic Modal Testing via Journal Bearing Lubricant Film – An Experimental Validation

Ilmar Ferreira Santos[✉] and Mads Nowak Frigaard

Department of Civil and Mechanical Engineering, Technical University of Denmark, 2800, Kgs. Lyngby, Denmark
ilsa@dtu.dk

Abstract. Experimental modal testing is performed in a rotor-bearing system using the journal-bearing lubricant film as an excitation source (shaker). The aerodynamic pressure field between the rotor and the bushing surface is influenced by the pressurized air injection through four orifices orthogonally machined on the bushing surface. The air injections are independently controlled by four piezoelectric stack actuators. Fluid film excitation forces resulting from the aerodynamic pressure variations are dependent on: i) the air supply pressures; ii) the DC amplitudes of the input signals to the piezoelectric actuators, i.e., opening-closing the flow through the orifices; iii) the AC amplitudes of the input signal – a Gaussian-distributed signal – supplied to the piezoelectric actuators; iv) different configuration of active injection orifice; and v) the journal angular velocity. The estimation of damping ratios and natural frequencies is done using a combination of four operational modal analysis (OMA) algorithms. Additionally, classical experimental modal analysis (EMA) is performed using an instrumented impact hammer and the results are used as a benchmark for the validation of the results coming from the novel technique combining the excitation via fluid film with output-only modal analysis algorithms. Good agreements between the two experimental approaches are found. The influence of the amplitudes of the input signal to the piezoelectric actuators, the air supply pressure, the orifice activation layouts, and the rotor angular velocities on the natural frequencies and damping ratios of the rotor-bearing system are investigated using both experimental dynamic testing techniques.

Keywords: rotordynamic testing · experimental modal analysis · output-only modal analysis · air bearings · active lubrication

1 Introduction

Conventional machine elements like bearings, brakes, clutches, couplings, and seals among others, are being transformed into smart mechatronic devices through the integration of sensing and actuation systems, i.e., electromechanics. Due to the capability of storing a large amount of measurement data from smaller, cheaper, and more reliable sensors as well as performing faster computational calculations with the data stored, the use of health condition monitoring techniques in rotordynamics has been significantly expanded in the last decades, especially the last one driven by the Industry 4.0 framework.

Active Magnetic Bearings (AMBs) are smart mechatronic devices commonly used in industrial applications to levitate rotors and control and stabilize their movements, but also to excite and generate rotor movements in case of performing rotordynamic testing [1,2]. In comparison with AMBs, controllable fluid film bearings (CFFBs) are "young" smart mechatronic devices, designed in a close synergistic relationship with different types of electromechanical actuators [3,4]. For example, oil or water-lubricated bearings are normally connected to servo valves and hydraulic servosystems [5–10], while gas or air-lubricated bearings are linked to piezoelectric actuators and pneumatic systems [11–13]. The fluid film forces in CFFBs - as the electromagnetic forces in AMBs - can also be used to control as well as excite rotor vibrations [1,2,14]. While the electromagnetic forces result from the dynamic interaction between the electromagnetic fields of the rotor, the stator, and the coils, the fluid film forces result from the fluid pressure fields between the rotor and the bearing surface, and they are dictated by the lubrication regimes, i.e., hydrodynamic, hydrostatic, and controllable, in the case of liquid lubricants, or aerodynamic, aerostatic, and controllable in the case of gas/air.

AMBs can generate "calibrated" forces to aid rotordynamic testing, taking advantage of their sensing and actuating systems to control the electromagnetic field and build small perturbation forces. One of the first contributions to the topic was given by Ulbrich [1] over three decades ago, Nowadays, the performance of rotordynamic testing via AMBs is a reality for the commissioning of many large industrial re-injection compressors [15]. CFFBs can also aid the performance of rotordynamic testing if the aerodynamic or hydrodynamic pressure fields can be influenced and controlled, aiming at building small "noninvasive" perturbation forces. The first contribution to the topic was given by Santos and Cerda in [16,17] about one decade ago. With "noninvasive" perturbation forces, the authors meant forces capable of perturbing the rotor-bearing system via fluid film without significantly changing the original static equilibrium position of the journal, i.e., without altering the rotor-bearing system dynamics [14]. Nevertheless, building a "noninvasive calibrated shaker" via the lubricant film is not a trivial task, once several transfer functions among mechanical, hydraulic, and electronic components are necessary [16,17] and the fluid film forces resulting from the aerodynamic or hydrodynamic pressure fields are strongly dependent on the rotor angular velocity and the fluid flow regime, in contrast to AMBs and the electromagnetic forces. One way of bypassing the necessity of knowing all these transfer functions is by using output-only modal analysis (OMA) techniques [14].

In this framework, the paper gives an original experimental contribution to the field of rotordynamic testing using a novel noninvasive modal testing procedure which combines controllable lubrication and output-only modal analysis techniques. A test rig composed of a flexible rotor supported by a controllable air bearing is used to validate the novel procedure. Piezoelectric actuators are used to control the radial air injection and the generation of the perturbation fluid film forces.

2 Test Apparatus and Experimental Procedure

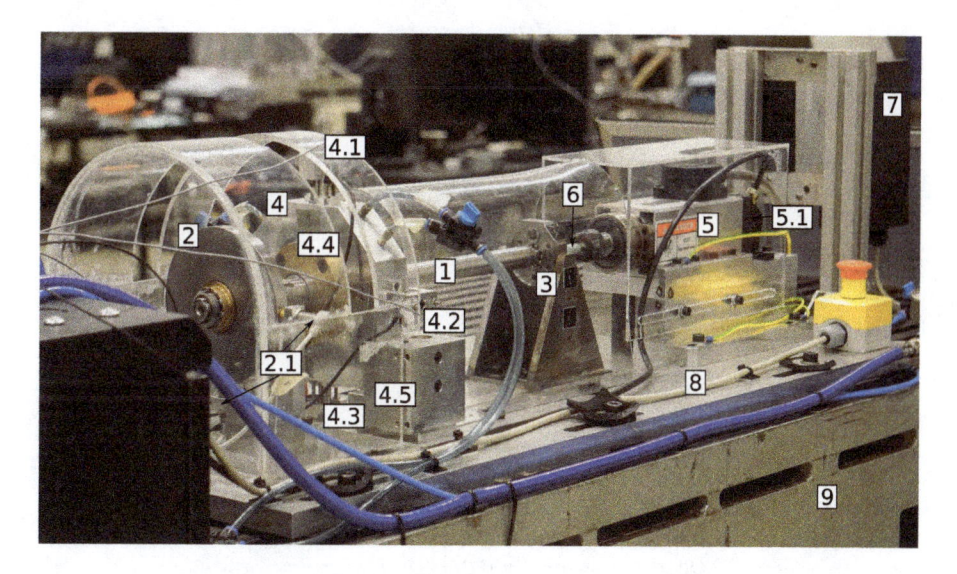

Fig. 1. Photo of the test rig illustrating the individual components of the test rig: **1** – flexible steel shaft, **2** – overhung steel disc, **2.1** – horizontal and vertical displacement sensors, **3** self-aligning ball bearing, **4** active gas bearing, **4.1** piezoelectric actuator 1, **4.2** – piezoelectric actuator 2, **4.3** – piezoactuator 3, **4.4** – bronze bushing, **4.5** – adjustable piezoelectric actuator mounting, **5** – AC motor, **5.1** – incremental encoder, **6** – flexible coupling, **7** – VLT micro drive, **8** – 20 mm aluminium plate, and **9** steel foundation.

The experimental campaign is carried out using the test rig seen in Fig. 1. The test rig is composed of a steel shaft **1** connected to an AC motor through a flexible coupling. A steel overhung disc **2** is mounted at the extremity of the steel shaft, and **1** plus **2** are referred to as the rotor. The rotor is supported by a self-aligning bearing **3** and an actively-controlled gas bearing **4**. The active gas bearing consists of a bronze bushing **4.5** mounted in an aluminum housing. Four piezoelectric stack actuators, **4.1**–**4.3**, are connected to the housing and control the air flow injection into the bearing gap. The actuators are orthogonally mounted around the housing, 1 and 3 in the vertical direction and 2 and 4 (not shown in the figure) in the horizontal direction. The gas bearing is supplied by pressurised air between 3 and 7 bar. The AC motor **5** is controlled by a VLT microdrive **7**. The rotor angular position and velocity are measured using an incremental encoder **5.1**. The lateral displacements of the rotor are captured at the disc location using two eddy current sensors orthogonally installed in the horizontal and vertical directions. The piezoelectric system is used for controlling the airflow into the gap at four locations, allowing for different configurations and strategies while generating time-dependent variations of the aerodynamic pressure field, to build the noninvasive fluid film forces to aid the performance of modal testing based on OMA. Table 1 presents the main dimensions and geometric properties of the rotor-bearing system.

Table 1. Test rig parameters.

Shaft length	500	mm
Shaft mass	2.04	kg
Disc diameter	139	mm
Disc thickness	13	mm
Disc mass	1.5	kg
Bearing diameter	40	mm
Bearing clearance	35.1 ± 1.02	μm
Bearing axial length	40	mm
Number of injectors	4	–
Injection nozzle radius	1	mm
Shaft material	Steel	–
Disc material	Steel	–
Bearing bushing material	Bronze	–

The experiments are designed based on multiple 2^k factorial designs [19], all replicated four times. The parameters studied were: **A** the angular velocity, **B** injection pressure, **C** initial position of the piezoelectric actuator (DC-offset), **D** active injector configuration, **E** amplitude of the piezoelectric input signal (excitation amplitude controlled by the AC part of the signal), and **F** the number of valves used to excite the rotor-bearing system. The parameters with corresponding factor levels can be seen in Table 2. Note that **F** cannot be at a high level "+" when **D** is at a low level "−".

The time used exclusively on the experimental campaign, i.e., without signal post-processing via OMA and EMA algorithms, can be seen in Table 3. Gaussian-distributed signals are generated and used as the input signal for the piezoelectric stack actuators. The signals change the position of the piezoelectric restrictors, altering airflow and pressure to the air gap accordantly with the input signal. A small portion of the aerodynamic pressure field will behave according to the input signal, generating fluid film forces capable of exciting the rotor-bearing system lateral dynamics. The system response is captured by the displacement sensors and is used in combination with OMA algorithms to estimate the system modal

Table 2. The design of experiments – the model terms and factor levels. Valves 1 and 3 are vertical and 2 and 4 horizontal.

Model Term	A	B	C	D	E	F
Factor level	Angular Velocity [RPM]	Injection Pressure [bar]	DC-Offset [V]	Valves Open	Excitation Amplitude [V]	Valves Excited
−	1000	3	0	1,3	1	1
+	5000	7	5	1,2,3,4	2.5	1,2,4

parameters, i.e., its natural frequencies and damping ratios. OMA techniques are well suited for estimating the system modal parameters since the system is time-invariant, low-damped, and excited by Gaussian distributed excitation [18]. Four OMA algorithms used are: three time-domain algorithms, i.e., Ibrahim Time Domain (ITD), Eigensystem Realisation Algorithm (ERA), and Data-Driven Stochastic Subspace Identification (SSI), and one frequency-domain algorithm, i.e., Enhanced Frequency Domain Decomposition (FDD) [18].

Table 3. Time used on the experimental measurement campaign.

Test Type	OMA			EMA	
Experiments	001–032	033–128	129–192	001–016	017–064
Time used [h]	7	11	6	2	8

For all measurements, a sampling rate of $f_s = 2500\,\text{Hz}$ is utilized, and to ensure robust estimations of the modal parameters, the tests are performed in intervals of 120 s. The raw data from the two displacement sensors are "run-out compensated", aiming at removing the unbalance response and harmonics not correlated to the excitation forces. The "run-out compensation" is implemented by conducting a 10 s pre-experiment where the rotor-bearing system is set to the desired operating condition and data is collected without the contribution of the fluid film force excitations, i.e., no input to the piezoelectric actuators. In the final experiment, the fluid film excitation forces are generated via actuators' input signals, and the data of the pre-experiment is subtracted from the final testing data, taking advantage of the rotor angular position obtained from the incremental encoder, used as a trigger (time reference) to correctly subtract one signal from the other. An averaging of the displacement data by each angle is done, and the mean of the run-out data is subtracted from the experimental displacement data. The encoder is mounted on the motor housing, and due to the flexible coupling, the run-out data and the raw experimental data is phase shifted and have to be phase-compensated. This is done before subtracting the run-out data from the experimental data. This is carried out as follows: i) averaging the displacement data by each angle for the run-out and experimental data, ii) shifting the run-out data, either forward or backward, to obtain the lowest root mean square error between the two signals. An experimental example can be seen in Fig. 2. When the run-out data has been correctly shifted, the subtraction of the run-out data and the raw experimental data can be completed.

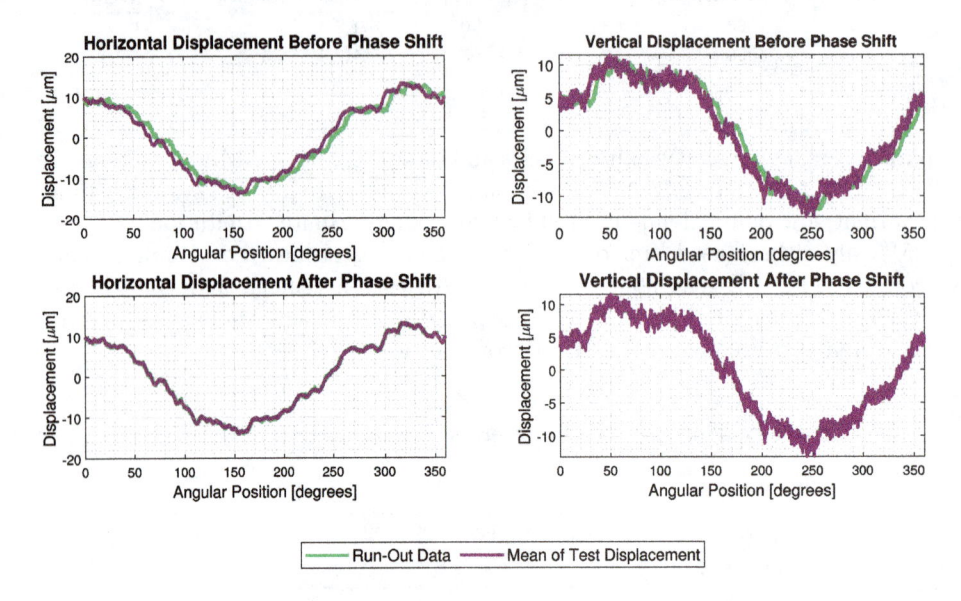

Fig. 2. The run-out compensation being shifted, $-7°$ to align with the test data, OMA Test - 001

The "run-out compensated" data is used to calculate the correlation function using a noise tale window to extract the non-decayed part of the correlation function. The correlation function is inputted into the ITD and ERA algorithms, and the run-out compensated data is fed directly into the SSI algorithm. To achieve better estimates, oversizing is needed since ITD, ERA, and SSI do not include noise modeling in estimating the modal parameters. This is especially important for dealing with the remnants of the unbalance response and electric noise. The spectral density function is calculated as a function of frequency using the Welch averaging technique with a Hanning Window and 50% overlap. The spectral density function is fed to the FDD and is transformed into singular values. The singular values are used to identify the system's natural frequencies by observing amplifications in the values. A frequency band is selected around observed natural frequencies and is inverse Fourier transformed back into the time domain. The free decay obtained is subsequently fed into the ITD algorithm to obtain damping ratios.

3 Experimental Results

Figure 3 illustrates the natural frequency and damping ratio for the highest participation mode obtained via OMA with a 95% confidence interval, as a function of **A** the rotor angular velocity, **B** the air injection supply pressure, **E** the amplitude of excitation defined by the amplitude AC of the signal fed to the piezoelectric actuators, and **D** the configuration of active orifices used to generate the excitation fluid film forces. The same natural frequency and damping ratio are also shown when EMA techniques with the instrumented impact hammer are employed. Figure 3 summarizes the results obtained during the whole experimental campaign and is very dense in terms of information. The first row illustrates the behavior of the rotor-bearing system's natural frequency and the second row the associated damping ratio. There are 4 blocks of figures per row, all of them showing the results obtained at two constant angular velocities, i.e., 1,000 rpm and 5,000 rpm. The first two blocks of figures on the top and the left-hand side illustrate the cases where the two orifices 1 and 3 are opened and the air supply pressure is kept at two different values, i.e., 3 bar and 7 bar. It is important to start evaluating the results using the red diamonds, which inform the results coming from the classical modal analysis and which are used as the reference results for comparisons with the results coming from the new modal technique combined with OMA algorithms. These two blocks of figures show that the system's natural frequency does not change much, because 1,000 rpm and 5,000 rpm are low angular velocities considering air bearings. All four OMA algorithms represented by four different colors - green (ITD), purple (ERA), yellow (SSI), and blue (FDD) - lead to similar and accurate values of natural frequencies. When the supply pressure is altered from 3 to 7 bar, a clear increase in the natural frequency is detected 118 Hz 135 Hz due to the pressurization of the bearing and an invasive action of the fluid film force changing the journal-bearing equilibrium position and the system dynamics. This conclusion can be supported by both modal testing techniques, the classical EMA and the one obtained using the lubricant film as a shaker combined with OMA. If we now evaluate the damping ratio associated with the natural frequency for the same conditions using the two blocks of figures in the second row on the left-hand side, we can also conclude that both modal techniques can accurately estimate the damping ratio. Both techniques show that when the level of pressurization changes from low (3 bar) to high (7 bar), a significant decrease in damping ratio can be detected, from approx. 6% (at 1,000 rpm) and 7% (at 5,000%) to 2% (at 1,000 and 5,000 rpm). This statement can be confirmed considering the two levels of the input signals used to feed the piezoelectric actuators and build fluid film perturbation forces of different magnitudes, i.e., 1 V and 2.5 V, meaning 10% and 25% of the maximum opening amplitude to the piezoelectric actuators.

Continuing with the analysis of the results of Fig. 3, the last two blocks of figures on the top and the right-hand side illustrate the cases where all four orifices 1, 2, 3, and 4 are opened and the air supply pressure is kept at two different values, i.e., 3 bar and 7 bar. Once again it is possible to conclude that both modal techniques lead to accurate results, comparing the red diamonds

(EMA) and the green, purple, yellow, and blue symbols, illustrating the outcomes from OMA algorithms. Of all four OMA algorithms, the ITD algorithm (green) is the one which leads to the largest confidence intervals. Generally, the sizes of the 95% confidence intervals are larger at high angular velocities, low excitation levels, and low supply pressure. It is worth mentioning that most of the outliers removed from Fig. 3 are estimations at a high angular velocity, low excitation amplitude, and low supply pressure. In terms of layout and configuration of active actuators, one can state that by increasing the number of valves exciting the rotor, one increases the accuracy of the natural frequency estimations and by increasing the injection supply pressure, smaller confidence intervals for natural frequency and damping ratio are achieved.

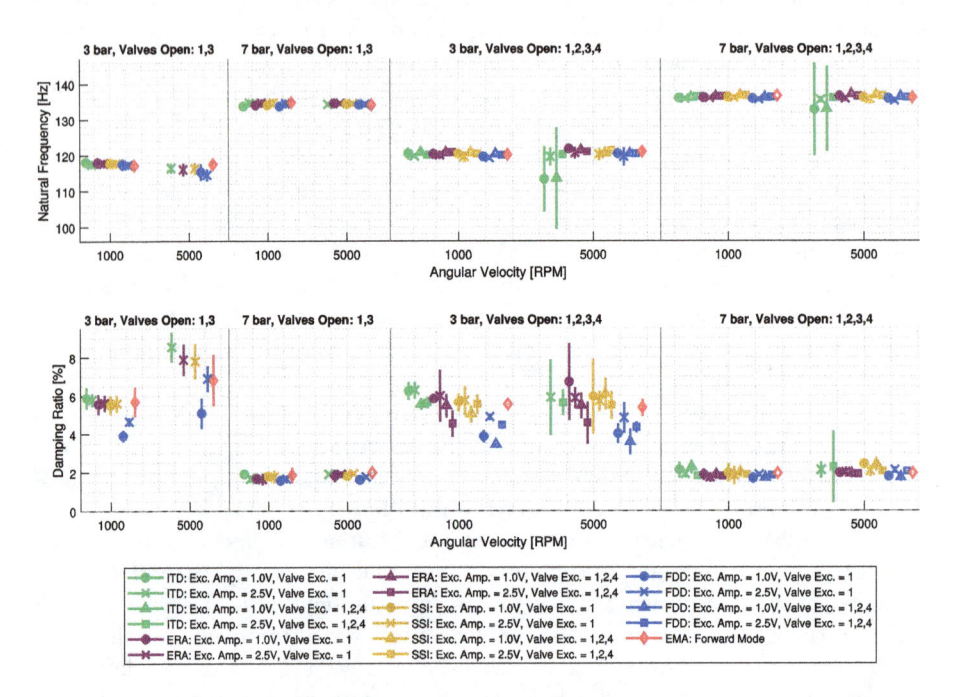

Fig. 3. Natural frequency and damping ratio estimates using EMA and OMA using several testing configurations, with a 95% confidence interval. Highest participation mode, with outliers removed.

For further analysis, the residuals between the 95% confidence interval of the OMA and EMA estimations are calculated. The residuals are determined as two parts, a lower and an upper bound, LB and UB, respectively. The lower bound is calculated by subtracting the upper 95% confidence interval for the EMA estimations from the lower 95% confidence interval for the OMA estimations. The upper bound is calculated by subtracting the lower 95% confidence interval for the EMA estimations from the upper 95% confidence interval for the OMA estimations. Hence resulting in the two worst-case scenarios for the 95% confidence interval. Table 4 contains the lower and upper bound for the natural frequency and damping ratio for all the OMA algorithms.

Table 4. Lower and upper bound of the 95% confidence interval residuals between EMA and OMA for the natural frequency and damping ratio at various test configurations. Highest participation mode.

Model Terms					Frequency [Hz]								Damping Ratio [%]							
					ITD		ERA		SSI		FDD		ITD		ERA		SSI		FDD	
A	B D	E	F		LB	UB	LB	UB	LB	UB	LB	UB	LB	UB	LB	UB	LB	UB	LB	UB
1000	3 (1,3)	1.0	(1)		−1.4	3.41	−1.64	3.36	−1.69	3.09	−2.26	2.96	−1.13	1.55	−1.44	1.17	−1.44	1.11	−2.85	−0.69
5000	3 (1,3)	1.0	(1)		−87.55	42.5	−38.3	20.65	−38.02	15.47	−5.48	1.22	−30.74	43.12	−9.4	13.25	−8.82	10.66	−3.85	0.42
1000	7 (1,3)	1.0	(1)		−1.94	−0.1	−1.66	0.2	−1.58	0.16	−1.96	−0.08	−0.44	0.58	−0.61	0.26	−0.52	0.41	−0.75	0.19
5000	7 (1,3)	1.0	(1)		−2.61	47.55	−1.25	2.09	−1.28	1.81	−1.54	1.79	−1.18	20.69	−0.49	0.16	−0.33	−0.01	−0.56	−0.18
Average:					−23.37	23.34	−10.71	6.58	−10.65	5.13	−2.81	1.47	−8.38	16.48	−2.98	3.71	−2.78	3.04	−2	−0.07
1000	3 (1,2,3,4)	1.0	(1)		−1.79	2.55	−1.36	1.67	−1.42	1.97	−1.94	0.87	0.22	1.2	0.11	0.48	−0.38	0.59	−2	−1.33
5000	3 (1,2,3,4)	1.0	(1)		−17.39	2.14	−0.8	2.4	−40.73	20.99	−1.6	0.55	38.32	53.32	−1.1	3.8	−1.82	2.99	−2.31	−0.39
1000	7 (1,2,3,4)	1.0	(1)		−1.57	0.35	−1.38	0.38	−1.31	0.53	−1.77	0.26	−0.23	0.6	−0.37	0.23	−0.45	0.5	−0.53	0.03
5000	7 (1,2,3,4)	1.0	(1)		−16.86	10.22	−0.96	1.73	−1.31	1.46	−1.53	1.04	−1.32	37.68	−0.2	0.23	0.35	0.6	−0.37	0.02
Average:					−9.4	3.81	−1.13	1.54	−11.19	6.24	−1.71	0.68	9.25	23.2	−0.39	1.19	−0.57	1.17	−1.3	−0.42
1000	3 (1,2,3,4)	1.0	(1,2,4)		−0.64	1.9	−0.56	1.58	−0.26	1.26	−0.56	0.95	−0.3	0.31	−0.72	0.57	−0.98	−0.03	−2.2	−1.99
5000	3 (1,2,3,4)	1.0	(1,2,4)		−22.2	7.37	−1.09	1.83	−1.74	0.81	−2.01	0.93	−3.96	80.8	−1.02	1.23	−0.55	2.01	−2.9	−0.65
1000	7 (1,2,3,4)	1.0	(1,2,4)		−0.67	−0.43	−0.61	0.17	−0.39	0.45	−0.88	−0.06	−0.03	0.67	−0.3	0.13	−0.19	0.31	−0.36	−0.1
5000	7 (1,2,3,4)	1.0	(1,2,4)		−15.59	9.29	0.18	1.58	−0.14	1.37	−0.5	0.91	−11.94	22.93	−0.25	0.18	0.1	0.79	−0.44	0
Average:					−9.77	4.53	−0.52	1.29	−0.63	0.97	−0.99	0.68	−4.06	26.18	−0.58	0.53	−0.41	0.77	−1.47	−0.68
Total Average, E = 1.0:					−14.18	10.56	−4.12	3.14	−7.49	4.11	−1.83	0.94	−1.06	21.95	−1.32	1.81	−1.25	1.66	−1.59	−0.39
1000	3 (1,3)	2.5	(1)		−1.77	2.64	−1.55	2.8	−1.72	2.78	−1.91	2.27	−1.01	1.24	−1.28	1.13	−1.34	1.15	−2.03	−0.04
5000	3 (1,3)	2.5	(1)		−3.44	1.31	−4.24	1.28	−3.86	1.53	−5.61	−0.31	−0.38	3.85	−1.08	3.22	−1.27	3.26	−1.91	2.1
1000	7 (1,3)	2.5	(1)		−1.69	1.46	−1.87	1.55	−1.76	1.66	−2.04	1.41	−0.76	0.4	−0.89	0.49	−0.8	0.65	−0.8	0.41
5000	7 (1,3)	2.5	(1)		−1.06	1.46	−0.76	1.64	−0.9	1.64	−1.18	1.52	−0.18	0.03	−0.27	0.09	−0.13	0.02	−0.35	−0.09
Average:					−1.99	1.72	−2.11	1.82	−2.06	1.9	−2.69	1.23	−0.58	1.38	−0.88	1.23	−0.89	1.27	−1.27	0.59
1000	3 (1,2,3,4)	2.5	(1)		−1.46	0.83	−1.76	1.68	−2.55	1.39	−2.03	0.36	0.28	1.22	−0.93	1.83	−0.58	0.98	−0.85	−0.46
5000	3 (1,2,3,4)	2.5	(1)		−4.55	1.84	−2.6	1.93	−3.08	1.6	−4.61	1.82	−1.88	2.97	−0.49	1.52	−0.66	1.34	−1.78	0.71
1000	7 (1,2,3,4)	2.5	(1)		−1.38	−0.13	−1.21	−0.06	−1.24	0.14	−1.64	−0.29	−0.23	0.18	−0.45	0.03	−0.59	0.38	−0.2	0.06
5000	7 (1,2,3,4)	2.5	(1)		−1.95	0.8	−1.65	1.01	−2.21	1.26	−2.1	0.53	−0.25	0.59	−0.22	0.31	−0.19	0.38	−0.02	0.33
Average:					−2.34	0.84	−1.81	1.14	−2.27	1.1	−2.59	0.6	−0.52	1.24	−0.52	0.92	−0.51	0.77	−0.71	0.16
1000	3 (1,2,3,4)	2.5	(1,2,4)		−1.26	1.2	−0.47	1.85	−0.45	1.05	−1.08	0.89	−0.24	0.42	−1.73	−0.3	−0.52	0.55	−1.3	−0.85
5000	3 (1,2,3,4)	2.5	(1,2,4)		−2.37	0.8	−1.46	1.37	−0.58	1.16	−1.49	0.41	−0.87	1.4	−2.35	0.74	−1.06	1.33	−1.76	−0.29
1000	7 (1,2,3,4)	2.5	(1,2,4)		−0.48	0.15	−0.6	0.28	−0.41	0.5	−0.68	−0.02	−0.25	−0.02	−0.28	−0.02	−0.23	0.1	−0.21	0.02
5000	7 (1,2,3,4)	2.5	(1,2,4)		−1.19	1.18	−0.43	1.17	−0.3	1.2	−0.66	0.8	−1.55	2.19	−0.17	0.07	−0.06	0.26	0.06	0.13
Average:					−1.32	0.83	−0.74	1.16	−0.44	0.98	−0.98	0.52	−0.72	1	−1.13	0.12	−0.47	0.56	−0.8	−0.25
Total Average, E = 2.5:					−1.88	1.13	−1.55	1.37	−1.59	1.33	−2.09	0.78	−0.61	1.21	−0.85	0.76	−0.62	0.87	−0.93	0.17
Total Average					−8.03	5.85	−2.83	2.26	−4.54	2.72	−1.96	0.86	−0.84	11.58	−1.08	1.28	−0.94	1.26	−1.26	−0.11

Table 4 shows that when the system is at high angular velocities, low excitation amplitudes, and with valves 1 and 3 opened, the upper and lower bound for the natural frequency and damping ratio estimations are significantly high for the ITD. The same can be observed for ERA and SSI at low injection pressure. By increasing the number of open valves, one decreases the spread of the bound for the natural frequency and the SSI damping ratio bound. The opposite effect can be observed for the natural frequency bound for the SSI, and ITD damping ratio bound, with an increase in the bounds. By increasing the number of valves exciting the system, one significantly increases the average accuracy of the SSI estimations and it has a small positive effect on the FDD and ERA estimations.

4 Conclusion and Future Aspects

After an extensive experimental campaign, the paper proves experimentally that fluid film forces coming from hydrodynamic and/or aerodynamic bearings can be controlled and used as an excitation source to aid rotordynamic modal testing. To bypass the necessity of knowing or calibrating the fluid film forces, output-only modal analysis techniques are combined with the fluid film perturbation forces.

The use of incremental encoders attached to the rotor or motor allows for run-out compensation and removals of harmonic components from the signals before OMA algorithms are used and the input data processed. The use of such hardware is of fundamental importance.

By increasing the excitation amplitude of the fluid film forces via input signal fed to the piezoelectric actuators, one generally increases the accuracy of the modal parameter estimations, i.e., natural frequency and damping ratio. The same tendency is observed when increasing the air supply pressure through the orifices machined in the bearing surface. By significantly increasing the injection supply pressure, one decreases the damping ratio of the rotor system and hence improves the conditions for using OMA algorithms since a core assumption when using OMA is sufficiently low damping. In such cases, the outcomes coming from classical EMA techniques and OMA combined with excitations via lubricant film are even more comparable and close.

Nevertheless, significant changes in the supply pressures from 3 bar to 7 bar also clearly illustrate the idea of "noninvasive" and "invasive" perturbation forces built via lubricant film. By changing the equilibrium position of the journal-bearing system through the high values of injection pressure, one also alters the dynamic characteristics of the system. Attention has to be paid when high values of supply pressure are used during modal testing.

By increasing the number of valves, or piezoelectric actuators, to induce rotor vibrations, one does not significantly affect the accuracy of the estimations at high excitation amplitudes.

Among the four OMA algorithms used, the overall most accurate is the FDD. It consistently has a small 95% confidence interval for the estimates of natural frequency and damping ratio.

In the future, a larger number of sensors will be installed on the test rig, aiming at also obtaining the mode shapes of the rotor-bearing system. Moreover, a larger frequency range will be used trying to identify additional natural frequencies and damping ratios.

References

1. Ulbrich, H.: New test techniques using active magnetic bearings, pp. 1011–1021 (1988)
2. Aenis, M., Knopf, E., Nordmann, R.: Active magnetic bearings for the identification and fault diagnosis in turbomachinery. Mechatronics **12**(8), 1011–1021 (2002)

3. Janocha, H.: Adaptronics and Smart Structures: Basics, Materials, Design and Applications. Springer, Heidelberg (1999). https://doi.org/10.1007/978-3-540-71967-0
4. Santos, I.F.: On the future of controllable fluid film bearings. Mech. Ind. **12**(4), 275–281 (2011)
5. Santos, I.F.: Trends in controllable oil film bearings. In: Gupta, K. (ed.) IUTAM Symposium on Emerging Trends in Rotor Dynamics. IUTAM Bookseries, vol. 1011, pp. 185–199. Springer, Dordrecht (2011). https://doi.org/10.1007/978-94-007-0020-8_17
6. Santos, I.F.: Design and Evaluation of Two Types of Active Tilting Pad Journal Bearings, pp. 79–87. Mechanical Engineering Publications, London (1994)
7. Bently, D.E., Grant, J.W., Hanifan, P.C.: Active controlled hydrostatic bearings for a new generation of machines, pp. 1–10, ASME Paper No. 2000-GT-0354 (2000)
8. Santos, I.F.: Controllable sliding bearings and controllable lubrication principles—an overview. Lubricants **6**(1), 16 (2018)
9. Salazar, J.G., Santos, I.F.: Feedback-controlled lubrication for reducing the lateral vibration of flexible rotors supported by tilting-pad journal bearings. Proc. Inst. Mech. Eng. Part J J. Eng. Tribol. **229**(10), 1264–1275 (2015)
10. Jensen, K.M., Santos, I.F.: Design of actively-controlled oil lubrication to reduce rotor-bearing-foundation coupled vibrations - theory & experiment. Proc. Inst. Mech. Eng. Part J J. Eng. Tribol. **236**(06), 1493–1510 (2022)
11. Morosi, S., Santos, I.F.: Experimental investigations of active air bearings, vol. 7 (2012)
12. Pierart, F., Santos, I.F.: Steady state characteristics of an adjustable hybrid gas bearing - computational fluid dynamics, modified Reynolds equation and experimental validation. Proc. Inst. Mech. Eng. Part J J. Eng. Tribol. **229**, 07 (2015)
13. Pierart, F.G., Santos, I.F.: Lateral vibration control of a flexible overcritical rotor via an active gas bearing - theoretical and experimental comparisons. J. Sound Vib. **383**, 20–34 (2016)
14. Santos, I.F., Svendsen, P.K.: Non-invasive parameter identification in rotordynamics via fluid film bearings: linking active lubrication and operational modal analysis. J. Eng. Gas Turbines Power **139**(06), 062507 (2017)
15. Almasi, A.: Modern high pressure gas injection centrifugal compressor for enhanced oil recovery. Oil Gas Eur. Mag. **37**, 196–199 (2011)
16. Santos, I.F., Cerda Varela, A.: Actively lubricated bearings applied as calibrated shakers to aid parameter identification in rotordynamics, vol. 7 (2013)
17. Varela, A.C., Santos, I.F.: Tilting-pad journal bearings with active lubrication applied as calibrated shakers: theory and experiment. J. Vib. Acoust. **136**(09), 061010 (2014)
18. Brincker, R., Ventura, C.E.: Introduction to Operational Modal Analysis. Wiley (2015)
19. Montgomery, D.C.: Design and Analysis of Experiments. Wiley (2017)

Problems of Development of Tribotronics: Control and Machine Learning Techniques

Leonid Savin[1], Denis Shutin[1(✉)], Yuri Kazakov[1], Alexey Kornaev[1], Roman Polyakov[1], and Shengbo Li[2]

[1] Orel State University n.a. I.S. Turgenev, Komsomolskaya str. 95, 302026 Orel, Russia
rover.ru@gmail.com

[2] School of Mechanical and Automotive Engineering, Xiamen University of Technology, Ligong Rd, 600, Jimei District, Xiamen 361024, China

Abstract. The main directions and problems of the development of tribotronics as an important component of a new technological stage are considered. Some technical solutions, such as active conical bearings and active fluid film dampers, are presented as challenging tribotronic systems. Simulation results show that friction can be considered as a controlled variable in addition to other parameters. The results also show that machine learning methods are well applicable in many typical design and calculation tasks of tribotronic bearings. Also a more detailed analysis of their applicability and the corresponding features is presented.

Keywords: Tribotronics · Fluid film bearings · Active conical bearings · Active dampers · Machine learning · Friction Control · Rotor Dynamics · Simulation

1 Introduction

The steady trend towards creating machines with higher power density and thermomechanical characteristics creates a need for new approaches to their design and calculation. Increasing requirements for energy efficiency and service life of machine elements also stimulate the development of tribological systems in the same direction. The result of these processes is the development of tribotronics as a field of science and technology, which is a convergence of mechatronics and tribology. Tribotronics is associated with integration of control and automated diagnostic functions in friction units using various actuators, sensors, controllers and computing devices. Tribotronic systems increase the efficiency, reliability, and service life of machines by controlling their power, kinematic, thermal, and geometric parameters. Consideration of machine elements as tribotronic systems makes it possible to reduce dissipative losses and provide the required integral and dynamic characteristics of moving machine parts by adding control loops to friction units. Bearings, dampers, seals, mechanical transmissions are the main objects of tribotronics in rotor machines. Considering them as tribotronic systems requires an increase in the complexity of simulation models used, in particular, integrating to them improved friction models. In [1], the influence of friction forces in the bearing on the accuracy of the pendular automatic rotor balance devices is considered. A new method for calculating the friction coefficient for journal bearings is presented in [2]. In [3] authors considered the calculation of the friction coefficient in bearings taking into account nonlinearities.

Bearings and dampers are usually the most loaded parts of rotor machines and also determine their resulting dynamic and energy parameters. Therefore, in them, first of all, means of active control of operating modes are introduced, which makes them tribotronic systems. Improvements made by control systems are usually related to vibrations and noise minimization, friction losses reduction, increase of reliability and service life [4, 5]. An overview of solution of active tribotronic fluid film bearings is made in [4] and [6]. The predominant number of works is devoted to bearings with adjustable gap geometry [4, 7–12].

This paper presents some problems and methods encountered within the framework of the tribotronic approach on the example of supports for new generation of rotary machines. Active conical fluid film bearings and active dampers are chosen as the demonstrative examples because their development includes solving complex problems of rotor dynamics and providing its stability, as well as because the authors of this research have experience in applying modern machine learning and control techniques to improve their performance.

Despite the typical problems with ensuring the stable rotor motion in conical fluid film bearings, there is a lack of studies relating to their adjustable design. Even in recent years most of conical bearings described in studies have convenient passive design. In [13] partial texturing of conical bearings increased their load capacity and reduced friction, in [14] an increase in stiffness and damping was achieved by texturizing the bearing surface in combination with using magnetohydrodynamic lubricant. The authors of this work also went from studying the characteristics of passive conical bearings [15, 16] to active correction schemes for their operation mode, the issues of controlling the position of the rotor and friction in them were considered in [17].

Active dampers are also relatively rarely considered in studies. A new non-linear controllable damper in which the rotor oscillation amplitudes are reduced using a smart material thickening fluid is presented in [18]. In [19] a controlled gap geometry allowed to adjust the damping properties of a squeeze film damper in real time. [20] proposes a new compact rotary magnetorheological damper with variable damping and stiffness being adjusted by an adaptive fuzzy system based on an artificial neural network (ANN). As in the case of conical bearings, the authors of this work have a background in the development of passive dampers [21, 22], as well as equipping them with control systems [23].

Thus, the paper presents some achievements in improving performance of active conical fluid film bearings and active dampers and by applying the relevant calculation and control techniques. The work demonstrates the ability of improving their dynamic and energy characteristics after adding the control facilities and considering them as complex tribotronic systems. Applicability of both intellectual and rather simple linear approaches to controlling their operation is illustrated by the obtained simulation results. Finally, an analysis of application of machine learning methods to some typical tasks in the field of tribotronic bearings is given with some conclusions about upcoming challenges in this area.

2 Tribotronics in Rotary Machines: Problems and Solutions

2.1 Active Conical Bearings

Conical bearings provide both axial and radial load capacity and eliminate the need in using separate radial and thrust bearings. However, it is much more difficult to ensure stable rotor motion in conical supports, as well as to tune the rotor system in general. Integration of active control means allows to control rotor motion and also other system parameters, including friction in the lubricant film. Some results relating to active conical bearings will be presented below.

Rotor Position Control in Active Conical Bearings

The thickness of the lubricant film is one of the main parameters of hydrodynamic bearings. The load capacity increases with a decrease in film thickness. The geometry of fluid film in conical bearings is described also with the rotor eccentricity and the average clearance value. The latter can be adjusted by axial displacement of the conical bearing. The proposed adjustable system includes a rotor supported by two oil- or water-lubricated conical bearings and a control system (see Fig. 1).

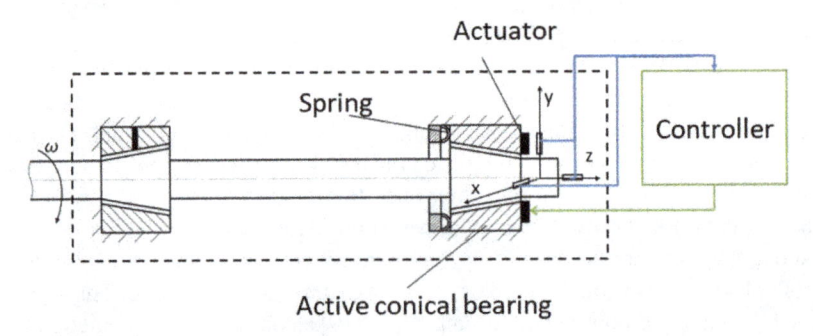

Fig. 1. Rotor on active conical bearing with adjustable average clearance

One of the bearings in the considered system is active due to the axial movability of its sleeve. The position adjustment is provided by the combination of the actuator and the elastic element (spring) at the opposite side. Displacement sensors are used to implement the feedback on the position. The controller obtains signals from the sensors and generates the necessary control action by changing the signal on the actuators. The controller can be built using either classical control principles or artificial intelligence methods.

A simulation model of the described system was developed. The mathematical model of fluid flow is based on the equation of motion and the equation of continuity of the medium:

$$\rho \frac{d\vec{v}}{dt} = \nabla \cdot T_\sigma + \rho \vec{f}, \tag{1}$$

$$\frac{\partial \rho}{\partial t} + \nabla \cdot (\rho \vec{v}) = 0. \tag{2}$$

where T_σ is a stress tensor, $\vec{v}$ is fluid flow velocity vector, ∇ is Hamiltonian, ρ is fluid density, $\vec{f}$ is a mass force.

Application of the Reynolds method for integrating Eq. (1) and Eq. (2) leads to the generalized Reynolds equation in the form [17]:

$$h^3 \frac{\partial}{\partial \beta_1}\left(\beta_1 \frac{\partial p}{\partial \beta_1}\right) + \frac{1}{\beta_1}\frac{\partial}{\partial \beta_2}\left(h^3 \frac{\partial p}{\partial \beta_2}\right) = \mu u_1 6h + \mu u_2 6 \frac{\partial h}{\partial \beta_2} - \mu 12 u_3 \beta_1. \quad (3)$$

where $u_1 = V_3$, $u_2 = \omega\left(r_0 + \frac{\beta_1 - \beta_1^-}{l_3}(r_1 - r_0)\right) + V_1 \cos\left(\frac{\beta_1^- \beta_2}{r_0}\right) - V_2 \sin\left(\frac{\beta_1^- \beta_2}{r_0}\right)$, $u_3 = V_1 \sin\left(\frac{\beta_1^- \beta_2}{r_0}\right) + V_2 \cos\left(\frac{\beta_1^- \beta_2}{r_0}\right)$ are the components of the fluid velocity vector at the shaft surface, $\beta_1, \beta_2, \beta_3$ are the angular, radial, and axial coordinates, respectively, p is pressure, μ is viscosity, h is a gap. The fluid film forces are calculated by solving the Reynolds equation. Finally, the rotor trajectories are obtained by calculating them iteratively together with the rotor dynamics equations.

The simulated rotor trajectories and viscous friction torque at different sleeve positions are presented in Fig. 2 and Fig. 3 respectively.

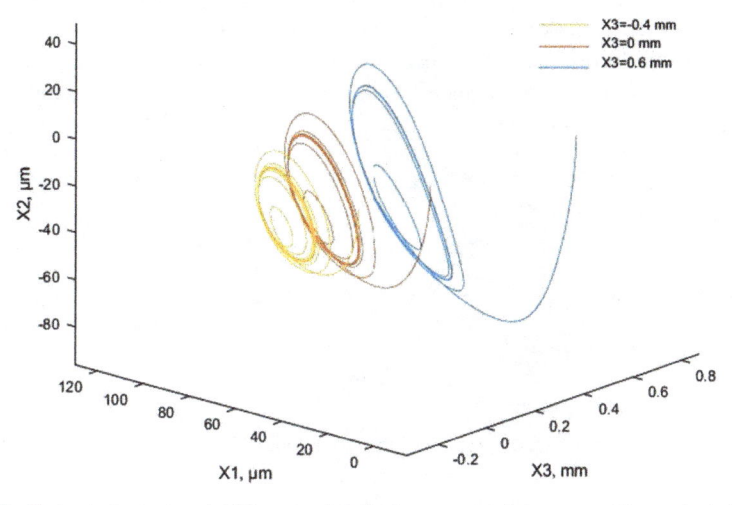

Fig. 2. Rotor trajectories at different axial displacements of the moveable conical sleeve

With an increase in the average clearance, the friction torque in the bearing decreases. However, this also leads to a decrease in load capacity, and the amplitude of the rotor oscillations increase. This effect should be taken into account in control system, which encourages choosing more modern control systems based on artificial intelligence.

Friction Control in Actively Lubricated Conical Bearings

Figure 2 shows the decrease in the rotor oscillation amplitudes with a decrease in the average clearance in the active bearing. The impact on the lubrication mode and the rotor motion inevitably leads to a change in the tribological parameters of the system. The

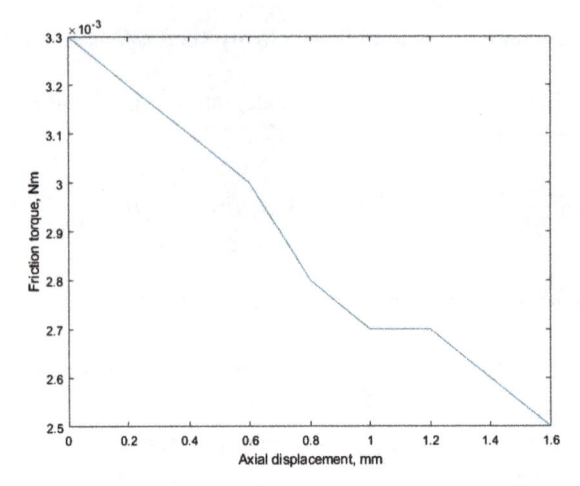

Fig. 3. Friction torque at different axial displacements of the moveable conical sleeve

relationship between the rotor motion and the viscous friction value in the conical bearing was investigated for an actively lubricated conical bearing. Unlike the configuration shown in Fig. 1, in this case, the control force is applied to the rotor by changing the lubricant supply pressure at the end of the sleeve, while the bearing elements are made stationary. A schematic of such a system is shown in Fig. 4, and a more detailed description of the study is given in [17].

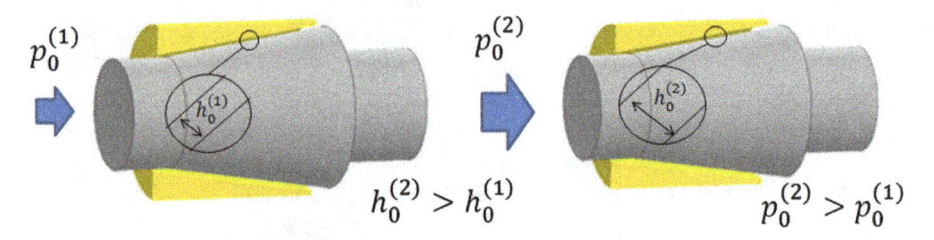

Fig. 4. Actively lubricated conical bearing

Active friction control in the described system was implemented with the feedback from a torque sensor mounted between the electric drive and the rotor. As described in [17], the best behavior of the system was demonstrated by the controller based on machine learning, namely deep Q-learning networks (DQN). Compared to traditional controllers, such as conventional and adaptive PID, the trained DQN controller provided not only a smooth change in the system state at the response to disturbances, but also effectively followed the imposed restrictions on the minimum clearance. The advantages in controlling friction and minimum bearing clearance are confirmed by additional computational tests, the results of which are shown in Fig. 5.

As can be seen from the results, the oscillation amplitude is hold stable even under changing load and meeting the constraints on the minimal allowable clearance. At the

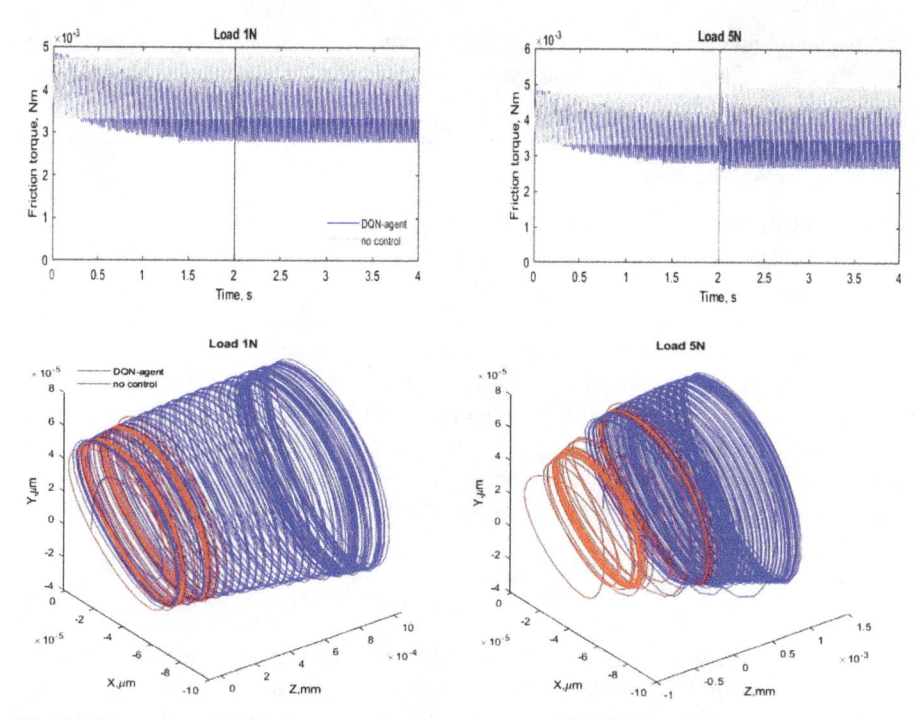

Fig. 5. Comparison of friction (top) and rotor trajectories (bottom) in passive and actively lubricated conical bearings at different loads applied (1 N and 5 N)

same time the viscous friction value is reduced by 15%, and also is hold constant under variable load. The severity of the friction reduction effect can also vary under different parameters and operating conditions of the rotor system, e.g., bearing, lubrication parameters, rotation speed, loads.

2.2 Performance Calculation of Rotor Motion in Fluid Film Bearings

Bearing reaction and friction torque can be represented as functions of the rotor position, rotation speed and lubricant supply pressure: $\vec{F}^b = \vec{F}^b(X_i, V_i, p_0)$, $M = M(X_i, V_i, p_0)$. It is known that ANNs provide non-linear approximation (interpolation) with high accuracy [24]. So, they can be used to improve mathematical models, primarily to speed up computationally expensive calculations, such as calculating the fluid film forces. The database for training ANNs is obtained by a single calculation of the system states using a physics-based simulation model.

The input parameters of the ANN for the approximation of the bearing forces will be the coordinates and velocities of the rotor in the bearing. The output parameters in this case will be the fluid film forces. After that, they can be calculated using the ANN approximation models within the set limits. The simulation results for evaluation the accuracy of approximation of the fluid film forces in the conical bearing are presented in Fig. 6.

The approximation error is less than 3%. Outside the training region, the shaft position error is up to 5%. However, the errors increase with increasing the axial displacement of the rotor. Comparative trajectories were obtained for 1 s of rotor motion. The simulation time using the numerical solution of the Reynolds equation was 631 s, and the simulation time using the ANN was 23 s, which is 27 times faster.

Advanced Control for Tribotronic Conical Bearings

The two considered designs of the controlled conical bearing demonstrate the two approaches to controlled tribotronic systems. In one case, a more conventional system with control of the spatial coordinates of a physical object (rotor) is considered. In the second case, the viscous friction in the bearing is considered as the controlled parameter, and data on the rotor position are used primarily to take into account the system constraints. Accordingly, one of the further tasks of tribotronics for such systems is to combine the above principles in a single system. In this case, the goal of control will be to achieve a minimum of a certain functional that combines both the rotor motion parameters, for example, oscillation amplitudes, and the friction parameters in the system. The control problem in this case can be represented as:

$$\mathbf{u} = \operatorname*{argmin}_{\mathbf{u} \in \vartheta} \sum_{k=1}^{\infty} \left[f(\Psi) + f(\Phi) \right], \tag{4}$$

where $\mathbf{u}$ is the vector of control signals, ϑ is the control vector subspace, Ψ and Φ are the vectors of spatial coordinates and friction parameters in the bearings, respectively. In this case, calculation of the necessary control is reduced to solving the Pareto-optimal problem. Being a multi-criterion problem, it can be also solved using the relevant machine learning methods, such as reinforcement learning, DDPG or DQN.

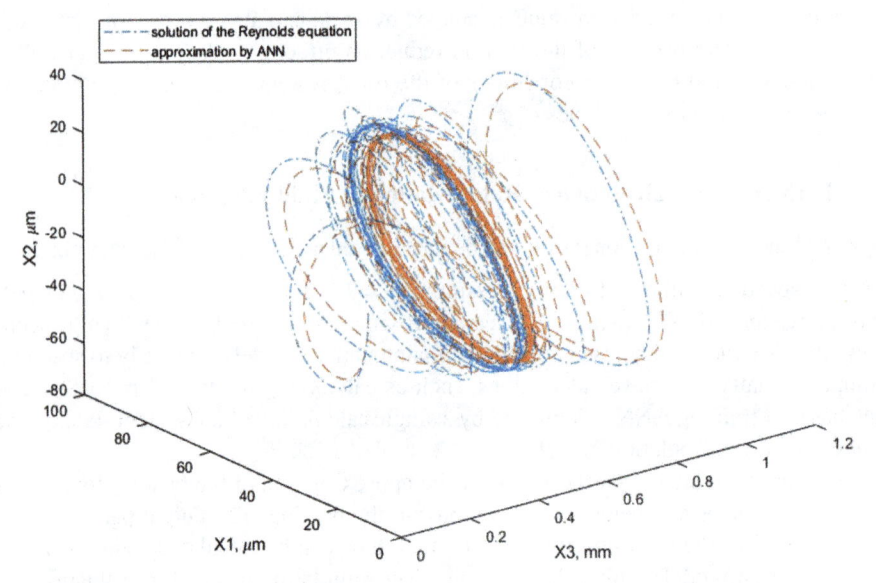

Fig. 6. Graph for comparing the results of solving the Reynolds equation

2.3 Active Dampers

Damping the rotor vibrations is usually effective only in resonant modes and for non-synchronous rotor oscillations with large amplitudes. In this case, the damper reduces the forces transmitted to the housing, but at certain values of the transfer coefficient, the operation becomes inefficient. So, adjustable dampers can improve the system behavior by decreasing the damping in the system after passing the resonance or leaving the zone of subsynchronous oscillations. The control system can be made either open-loop or closed-loop, with feedback on the analyzed parameters of the rotor motion. The benefits of active damping have been shown in some of the papers mentioned in the introduction. A fairly simple and efficient design of an active damper is presented in Fig. 7 [23]. It allows improve the dynamics of the rotor-bearings system significantly by adjustable damping properties.

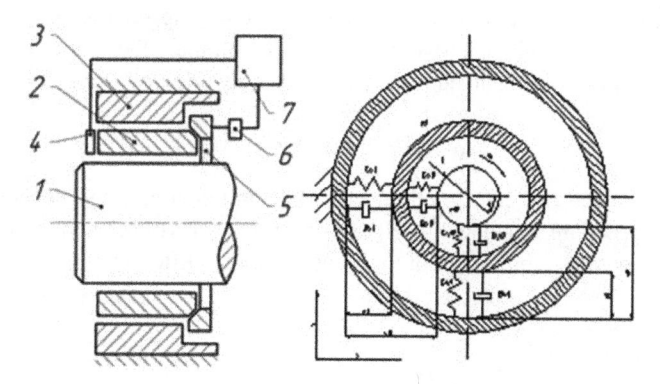

Fig. 7. Active bearing-damper system

The damper in Fig. 7 is switched on and off by the linear actuator 6. The elastic-damping bearing 1 also contains housing with a fixed 3 and a floating 2 sleeves. The motion of the clamping ring 5 is provided by the actuator 6 fixed at the housing. After starting the rotation, the bearing operates with the damper off. The shaft is carried by the bearing fixed by the clamping ring 5 in the housing. When a certain rotation speed and/or oscillation amplitude detected by a sensor 4 is reached, the damper is switched on by moving the clamping ring 5. With a decrease in the rotation speed and/or oscillation amplitude, all processes occur in the reverse order.

The functioning of this support in resonant modes with an active hydrodynamic damper (the clamping ring 5 is retracted) corresponds to the physical principles of a fluid film bearing with a floating sleeve to a certain extent. The simulation of the rotor motion in this case is based on the calculation of the pressure distribution and the dynamic stiffness and damping coefficients of the fluid films. The rotor trajectories are calculated as described for the conical bearing, with additional equation describing the thermal properties and variable surfaces geometry [25]. The dynamic model of the support is a two-mass two-power oscillator. The motion equations of its elements are as follows:

$$\begin{aligned} m_1\ddot{x}_1 &= -C_{x1}x_1 + C_{x2}(x_2 - x_1) - B_{x1}\dot{x}_1 + B_{x2}(\dot{x}_2 - \dot{x}_1); \\ m_2\ddot{x}_x &= -C_{x2}(x_2 - x_1) - B_{x2}(\dot{x}_2 - \dot{x}_1) + \Delta\omega^2 m_2 \cos(\omega t); \\ m_1\ddot{y}_1 &= -C_{y1}y_1 + C_{y2}(y_2 - y_1) - B_{y1}\dot{y}_1 + B_{y2}(\dot{y}_2 - \dot{y}_1) - gm_1 \\ m_2\ddot{y}_2 &= -C_{y2}(y_2 - y_1) - B_{y2}(\dot{y}_2 - \dot{y}_1) + \Delta\omega^2 m_2 \sin(\omega t) - gm_2, \end{aligned} \quad (5)$$

where x_1, y_1 are the displacements of the damper sleeve in the housing; x_2, y_2 are the displacements of the rotor relative to the housing; C_1, C_2 are stiffness coefficients of the damper and the lubricant film of the bearing; B_1, B_2 are damping coefficients of the damper and the lubricant film of the bearing; m_1, m_2 are the damper sleeve and rotor masses respectively; ω is rotor angular speed; t is time.

The integration of pressure distribution allows determining the fluid film forces in the bearing and the damper. The stiffness and damping coefficients are generally determined with the small perturbation method. To simulate the motion of the rotor in transient modes, Eq. (5) is solved together with the equation of moments:

$$J\varepsilon = M_d - M_c. \quad (6)$$

The rotor trajectories calculated under various initial conditions, such as the initial position and speed, are shown in Fig. 8 [23]. The trajectories of the rotor with the damper switched on indicate stabilization of the rotor motion in different cases.

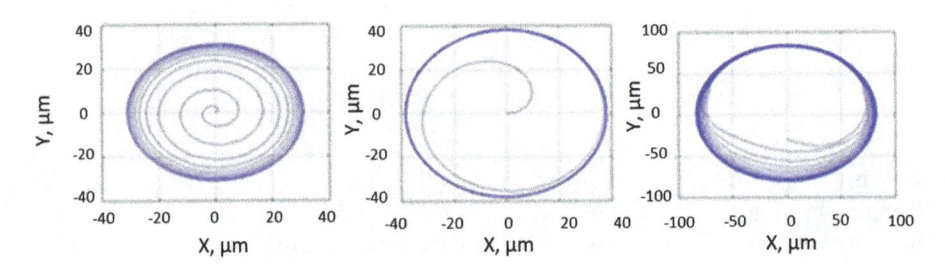

Fig. 8. Rotor trajectories under different initial conditions

Figure 9 shows the system operation under realistic conditions with stochastic disturbances. Despite the partly chaotic motion, the rotor orbit always moves to the constant limit cycle, proving the stable machine operation.

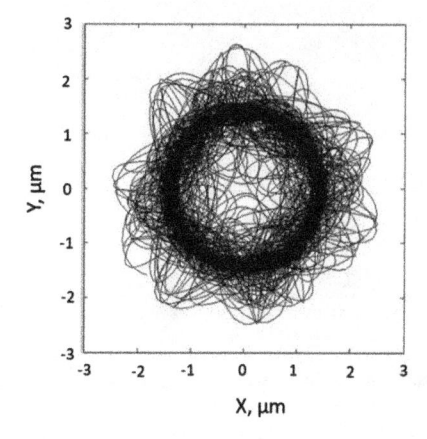

Fig. 9. Rotor trajectory under realistic conditions

The algorithm of operation of the active bearing-damper support is based on switching on and off the damper at certain moments when the rotor is accelerating or stopping. The examples of the spectrum of vibration displacements for a system with an active damper turned off and on are shown in Fig. 10, where the decrease in amplitude more than 3 times is observed.

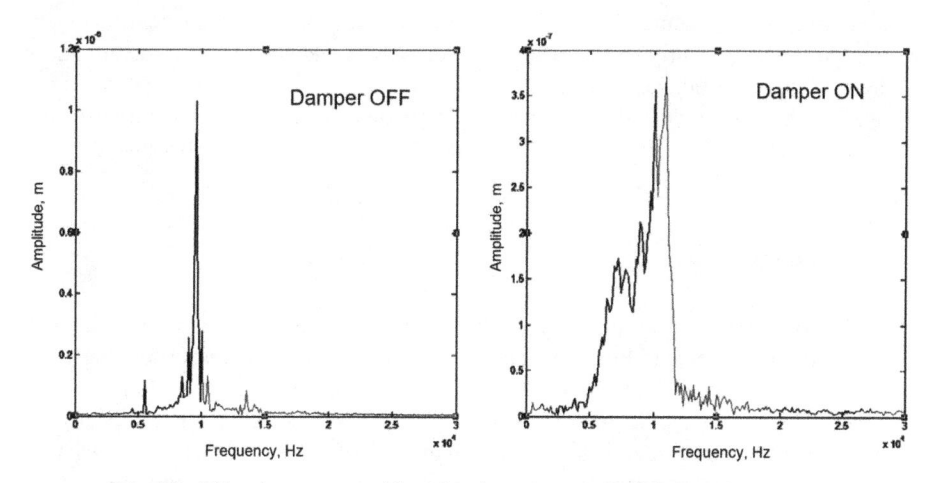

Fig. 10. Vibration spectra with active damper turned off (left) and on (right)

The basic controller implements discrete switching of the modes and considers the amplitudes of oscillations h_0 as a threshold. Figure 11 shows the results of simulation of the rotor-bearing system with the active damper during acceleration, where $\bar{y}$ is dimensionless vertical amplitude of the rotor oscillation, $\bar{n}$ is dimensionless rotational speed.

The resulting system response is a combination of responses of its two main states. The plot 1 in Fig. 11 represents the Bode plot of the system with the damper switched off,

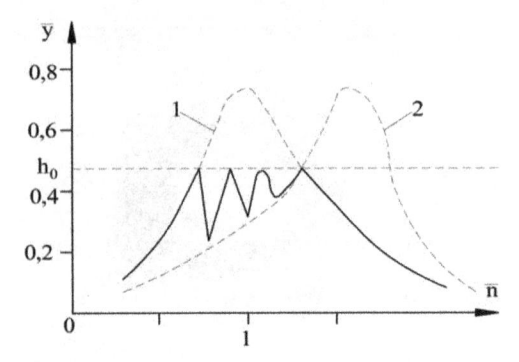

Fig. 11. Resulting vibration amplitudes during the rotor acceleration with the active damper

and the plot 2 – with the damper switched on. When approaching the resonant zone, the damping mode is switched on and the oscillation amplitudes decrease. With an increase in the rotation speed and a decrease in the amplitudes, the damping is turned off, which leads to a subsequent increase in the amplitudes, and further repetition of the scenario till passing the resonance.

The switching oscillation present in Fig. 11 can be eliminated by implementing a smooth control of the damping degree. Considering machine learning based controllers, DDPG method [26] is well applicable to such tasks due to its ability to deal with a continuous space of output signals. In this case, the advantage of using reinforcement learning is that it is a fairly simple method to consider the heterogeneity of disturbances in the system. In particular, it is possible to implement a differentiated response of the damper to the occurrence of nonlinear poly-harmonic oscillations in addition to the resonant vibrations, in particular, parametric oscillations associated with a change in the rotor transverse stiffness.

3 Discussion

The tribotronic approach to the design and calculation of rotor-bearing systems implies several significant aspects. First, the tribological processes in the corresponding units are considered as significant, and the friction modes are optimized to minimize the dissipative power losses. Secondly, in most cases supports have integrated means for adjusting their operation parameters, such as reduction of vibrations. In this case friction is also considered as a controlled parameter and is taken into account by controllers when calculating control signals. Since the direct measurement of friction parameters in elements of rotary machines is difficult in many cases, a model-based approach is often used. In this case, the friction value is calculated based on the values of other parameters that are directly measured.

Data-based methods are well applicable when working with tribotronic systems due to the non-linearity of processes, and also the need to simultaneously take into account many heterogeneous factors and parameters. In such conditions advanced computational methods are required, e.g. for solving the multi-criterion optimization problems when calculating the control signals taking into account both friction and the rotor position in bearings. Much attention to calculation and computational methods is also required to reduce the resulting computational cost of the problems being solved, e.g. model-based calculation of friction parameters. The criterion for the correct choice of the methods is an increase in the calculation speed by orders compared to physics-based models with comparable accuracy. As can be seen from the results presented in this study, such tasks in some cases are successfully handled by ANN-based approximators.

It should be noted that a real gain in time is obtained for cases where it is necessary to repeatedly implement calculations with the same simulation models and with the same assumptions and restrictions. Once receiving such data from a physics-based model, then after training it is possible to repeatedly use more productive tools. However, if any parameters of the simulated system should be changed, as well as assumptions and/or restrictions, the procedure for generating data and training the model needs to be repeated, which reduces the advantages of the approach.

Joint analysis of the typical tasks in the field of tribotronic bearings and the known capabilities and limitations of machine learning methods can be made, also considering the results of their joint use mentioned above. As the result, the main areas of application of such methods in tribotronic rotor-bearing systems and their features are presented in Table 1.

Table 1. Analysis of application of machine learning methods for tribotronic rotor-bearing systems

Tasks	Applications	Advantages	Disadvantages
Approximation *(e.g., with feedforward ANN)*	Approximation of fluid film forces in active bearings; approximation of mechanical properties of damping elements in adjustable dampers	1. The ability of approximating many functions of many variables using one ANN 2. High interpolation accuracy 3. High calculation speed	1. Large number of training samples is required 2. Low extrapolation accuracy, the model is applicable only within the parameters range covered by training samples
Prediction of time series *(e.g., with non-linear auto regressive network with external input)*	Prediction of rotor orbits, including under changing lubrication conditions	1. High interpolation accuracy 2. High extrapolation accuracy which allows to predict the behavior of the object (rotor) outside the training samples range 3. High calculation speed	1. Large number of training samples is required 2. Only functions of one variable (time) can be predicted

(continued)

Table 1. (*continued*)

Tasks	Applications	Advantages	Disadvantages
Pattern recognition and classification (*e.g., with two-layer feedforward ANN*)	Recognition of a certain type of rotor oscillation (synchronous, sub-synchronous, aperiodic, parametric, etc.)	1. High calculation speed 2. High accuracy is achievable – identifying distinguishing features that are difficult to establish in other ways; accuracy increases significantly with increase in a number of observed parameters	1. Large number of training samples is required
Control (*e.g., with reinforcement learning*)	Stabilizing rotor position at a setpoint; minimizing viscous friction in bearing	1. Reinforcement learning models do not require knowing a way to achieve the desired goal 2. The obtained solutions can be difficult to achieve otherwise	1. Large number of training attempts is usually required 2. Large number of hyperparameters to be tuned

The results of the analysis showed that in each case the use of machine learning methods in tribotronic systems is associated with several advantages and disadvantages. There are both specific and general features for the considered types of problems solved. The need for a large number of training samples is a common disadvantage for the most of tasks. However, in most cases a high calculation speed is achieved if training is implemented successfully.

In any case, machine learning methods should be considered as a productive tool for solving the current and future problems of tribotronics. They can be used both as an alternative and as a supplement to more conventional methods. In particular, the considered problems of motion and friction control in the "rotor - conical bearings", "rotor-bearing-damper" systems are the illustrative examples where the use of tribotronic approach, as well as machine learning tools, is justified.

4 Conclusion

Consideration of active rotor-bearing systems as tribotronic ones implies an extended formulation of problems at the stages of their design and operation. In particular, friction in elements of rotary machines and the corresponding power losses are considered as the controllable parameters in addition to the common tasks, such as vibrations reduction. It requires solving a number of new tasks related to the optimization of operation modes. In particular, it can be implemented by means of active control in the tribotronic devices. For tribotronic dampers the task is modified from minimizing vibrations in all operating modes to optimizing the function of the resulting amplitudes from operating parameters. For active conical bearings the problem of minimizing viscous friction is solved together with the problem of stabilizing the rotor motion.

The increasing complexity of the problems solved requires the use of new computational and calculation methods. In particular, machine learning methods are of interest for solving a number of applied problems of tribotronics. They can be used both instead of and in conjunction with more conventional methods. At the same time, the applicability of machine learning is not absolute due to the requirements for the amount and quality of training data. Thus, one of the urgent problems in the field of modern tribotronic systems is the choice of the most relevant design and calculation methods, as well as the algorithmic base for the control systems. Among other things, this choice consists in finding the optimal balance between the use of data-based approaches and the modeling of physical processes. The final decision should be made considering the properties of the specific tribotronic objects, the formulation of synthesis and analysis problems, and also the limitations of the methods themselves.

Acknowledgements. The study was supported by the Russian Science Foundation grant No. 22-19-00789, https://rscf.ru/en/project/22-19-00789/.

References

1. Ziyakaev, G.R., Gorbenko, M.V., Gorbenko, T.I., Ivkina, O.P.: Friction influence on the accuracy of the rotors automatic balance. Key Eng. Mater. **685**, 441–444 (2016). https://doi.org/10.4028/WWW.SCIENTIFIC.NET/KEM.685.441
2. Ünlü, B.S., Atik, E.: Determination of friction coefficient in journal bearings. Mater. Des. **28**, 973–977 (2007). https://doi.org/10.1016/J.MATDES.2005.09.022
3. Kumar, M., Whittaker, A.S., Constantinou, M.C.: Characterizing friction in sliding isolation bearings. Earthq. Eng. Struct. Dyn. **44**, 1409–1425 (2015). https://doi.org/10.1002/EQE.2524
4. Breńkacz, Ł., Witanowski, Ł., Drosińska-Komor, M., Szewczuk-Krypa, N.: Research and applications of active bearings: a state-of-the-art review. Mech. Syst. Signal Process. **151**, 107423 (2021). https://doi.org/10.1016/J.YMSSP.2020.107423
5. Qin, F., Li, Y., Qi, H., Ju, L.: Advances in compact manufacturing for shape and performance controllability of large-scale components-a review. Chinese J. Mech. Eng. **30**, 7–21 (2017). https://doi.org/10.3901/CJME.2016.1102.128
6. Santos, I.F.: Controllable sliding bearings and controllable lubrication principles—an overview. Lubricants **6**, 16 (2018). https://doi.org/10.3390/LUBRICANTS6010016
7. Chasalevris, A., Dohnal, F.: Enhancing stability of industrial turbines using adjustable partial arc bearings. J. Phys. Conf. Ser. **744**, 012152 (2016). https://doi.org/10.1088/1742-6596/744/1/012152
8. Chasalevris, A., Dohnal, F.: Modal interaction and vibration suppression in industrial turbines using adjustable journal bearings. J. Phys. Conf. Ser. **744**, 012156 (2016). https://doi.org/10.1088/1742-6596/744/1/012156

9. Chasalevris, A., Dohnal, F.: A journal bearing with variable geometry for the suppression of vibrations in rotating shafts: simulation, design, construction and experiment. Mech. Syst. Signal Process. **52–53**, 506–528 (2015). https://doi.org/10.1016/j.ymssp.2014.07.002
10. Chasalevris, A., Dohnal, F.: Improving stability and operation of turbine rotors using adjustable journal bearings. Tribol. Int. **104**, 369–382 (2016). https://doi.org/10.1016/J.TRIBOINT.2016.06.022
11. Pai, R., Parkins, D.W.: Performance characteristics of an innovative journal bearing with adjustable bearing elements. J. Tribol. **140** (2018). https://doi.org/10.1115/1.4039134/383984
12. Deckler, D., Veillette, R., Braun, M., Choy, F.: Simulation and control of an active tilting-pad journal bearing. Tribol. Trans. **47**, 440–458 (2010). https://doi.org/10.1080/05698190490463277
13. Shinde, A., Pawar, P., Shaikh, P., Wangikar, S., Salunkhe, S., Dhamgaye, V.: Experimental and numerical analysis of conical shape hydrodynamic journal bearing with partial texturing. Procedia Manuf. **20**, 300–310 (2018). https://doi.org/10.1016/j.promfg.2018.02.045
14. Singh, A., Sharma, S.C.: Behaviour of conical porous hybrid journal bearing operated with MHD lubricant considering influence of surface irregularities. Tribol. Int. **174**, 107730 (2022). https://doi.org/10.1016/J.TRIBOINT.2022.107730
15. Li, S.B., Ao, H.R., Jiang, H.Y., Korneev, A.Y., Savin, L.A.: Steady characteristics of the water-lubricated conical bearings. J. Donghua Univ. **29**, 115–122 (2012)
16. Koltsov, A.Y., Korneev, A.Y., Savin, L.A., Li, S.: Dynamic equilibrium surfaces for conical fluid-film bearings. IOP Conf. Ser. Mater. Sci. Eng. **233**, 012041 (2017). https://doi.org/10.1088/1757-899X/233/1/012041
17. Kazakov, Y.N., Kornaev, A.V., Shutin, D.V., Li, S., Savin, L.A.: Active fluid-film bearing with deep q-network agent-based control system. J. Tribol. **144**, 081803 (2022). https://doi.org/10.1115/1.4053776
18. Zhao, Q., Yuan, J., Jiang, H., Yao, H., Wen, B.: Vibration control of a rotor system by shear thickening fluid dampers. J. Sound Vib. **494**, 115883 (2021). https://doi.org/10.1016/J.JSV.2020.115883
19. Zheng, W., Pei, S., Zhang, Q., Hong, J.: Experimental and theoretical results of the performance of controllable clearance squeeze film damper on reducing the critical amplitude. Tribol. Int. **166**, 107155 (2022). https://doi.org/10.1016/J.TRIBOINT.2021.107155
20. Yu, J., Dong, X., Su, X., Qi, S.: Development and characterization of a novel rotary magnetorheological fluid damper with variable damping and stiffness. Mech. Syst. Signal Process. **165**, 108320 (2022). https://doi.org/10.1016/J.YMSSP.2021.108320
21. Sheng-Bo, L., Hui, Y., Hong-Yuan, J., Liang, C.: Analysis of dynamic performance of metal rubber damping ring applied in high-speed rotor system. Acta Phys. Sin. **61** (2012)
22. Komarov, M.A., Zhuihua, Z., Savin, L.A.: Questions of Projection of Bearing Nodes of Fluid-Flow Friction with Damping Elements. Proc. Oryol State Tech. Univ. Ser. Fundam. Appl. Probl. Eng. Technol. **271** (2008)
23. Savin, L.A., Bychkov, M.V., Kozyrev, D.L., Marakhin, N.A.: Trajectories of movement of rotors with controlled elastic-damper supports. Fundam. Appl. Probl. Eng. Technol. 105–111 (2020)
24. Kornaev, A.V., Kornaev, N.V., Kornaeva, E.P., Savin, L.A.: Application of artificial neural networks to calculation of oil film reaction forces and dynamics of rotors on journal bearings. Int. J. Rotating Mach. **2017** (2017). https://doi.org/10.1155/2017/9196701
25. Моделирование роторных систем с опорами жидкостного трения (2006) (5-94275-258-3) — книга автора Савин Л.А., Соломин О.В. | НЭБ, https://rusneb.ru/catalog/010003_000061_e623be03db0220589221cd1eec20347e/. Accessed 20 July 2022
26. Lillicrap, T.P., et al.: Continuous control with deep reinforcement learning. 4th Int. Conf. Learn. Represent. ICLR 2016 - Conf. Track Proc. (2016)

A Presentation of Control Theory Applied to the Design of Controllable Segmented Gas Foil Bearings

Janus Walentin Jensen and Ilmar Ferreira Santos[✉]

Department of Civil and Mechanical Engineering, Technical University of Denmark,
2800 Kgs. Lyngby, Denmark
{jwaje,ilsa}@dtu.dk
https://construct.dtu.dk/

Abstract. With industry diverging away from the use of oil in the hopes of a more environmentally friendly production, the use of Gas Foil Bearings (GFBs) can facilitate an oil-free alternative for high-speed rotating light machinery. Currently, their application is, however, limited by their low load-bearing capability and low vibration damping. Hybridization has the possibility to mitigate these limitations. The system presented in this paper demonstrates the increased system capabilities through passive and active Hybrid Gas Foil Bearings (HGFB) with radial gas injection. The non-linear differential equations comprising the system, pressure states, foil deflections, rotor movements, and actuator position, are solved simultaneously with and without control feedback. The presented results comprise unbalance response waterfall diagrams.

Keywords: Gas Foil Bearings · Injection · Control

1 Introduction

While the use of Gas Foil Bearings (GFBs) in lightly loaded high-speed rotating systems is widespread, their use in heavier-loaded machinery is limited by inherent physical limitations, among the most important is the poor dynamical stability and sub-synchronous vibrations, due to low damping, and low load bearing capacity, resulting in considerable wear at the start and stop conditions.

The need for oil-free technology has propelled the advancements in the mathematical modeling of the GFB, from the first analyses by Oh and Rohde [1], Heshmat et al. [2], and Peng and Carpino [3], to studies optimizing the bearing geometry [4] and the analysis of novel bearing designs [5].

To advance the bearing applicability in heavier machinery, various passive [6,7], regulated [8,9], and feedback-controlled systems [10,11] have been implemented. Although hybridization increases the complexity of the system with added subsystems, the advantages cannot be disregarded for continued advancements. The GFBs in this paper are hybridized through pressurized radial injection, shown to have a positive effect on the system by von Osmanski and Santos

[25]. The use of radial injection has proven its worth through use in several decades in rigid gas bearings as described in [12–17].

2 Design Configuration and Operation Modes

The present work is an augmented version of the GFB rotor system presented in a passive format by Larsen et al. [18] and later analyzed in multiple papers [19–24]. The system was expanded to a hybrid format by von Osmanski and Santos [25], and with actuator degrees of freedom added later by Heinemann [26].

2.1 Test Rig

The modeled system is a test setup that mimics a direct-driven turbo blower with a nominal speed range of $15-30$ kRPM. The system constitutes a 21.17 kg hollow rotor shaft with an approximate length of 0.59 m, as seen in Fig. 1. The shaft is supported by two duplicate GFBs, which are placed at either end of the rotor, with discs placed at the rotor extremity. The rotor is encased in a steel shell through which four inductive eddy-current proximity sensors are mounted, measuring the rotor's vertical and horizontal position, close to either bearing.

The relevant geometrical parameters can be seen in Fig. 1, with dimensions given in Table 1.

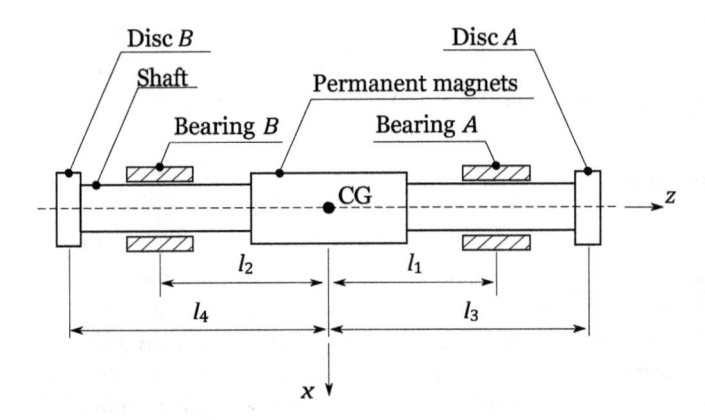

Fig. 1. Schematic of the test rig. Recreated from von Osmanski and Santos [25].

2.2 Gas Foil Bearing Configuration

The three-pad Hybrid Gas Foil Bearings (HGFBs) are based on second-generation Siemens GFBs [27] augmented with three high-pressure injectors using air as a lubricant. Each of the three equidistant, bearing pads consists of a top foil spanning 115°. The top foil is clamped in the leading inlet edge and free in the trailing edge. The inlet edge has an initial slope height, h_s. The top foil is supported by a corrugated foil, denoted bump foil, below which shims are placed to obtain the required bearing clearance C. The relevant geometrical parameters for the HGFBs can be seen in Fig. 2a, with dimensions and mechanical properties given in Table 1.

Three high-pressure gas injectors are placed at the center of the foils, as presented by von Osmanski and Santos [25].

The injectors consist of a pin, whose position is controlled by a piezoelectric stack actuator acting against a series of Belleville springs as illustrated in Fig. 2b. The pin acts as a cone restrictor, restricting the gas flow from the modified supply pressure to the lubricating film. The relevant geometrical parameters used in the model can be seen in Fig. 3. Figure 3b, additionally, serves as a sketch of the injector Computational Fluid Dynamics (CFD) model used by von Osmanski and Santos [25]. The dimensions and mechanical properties used for the injector are listed in Table 2.

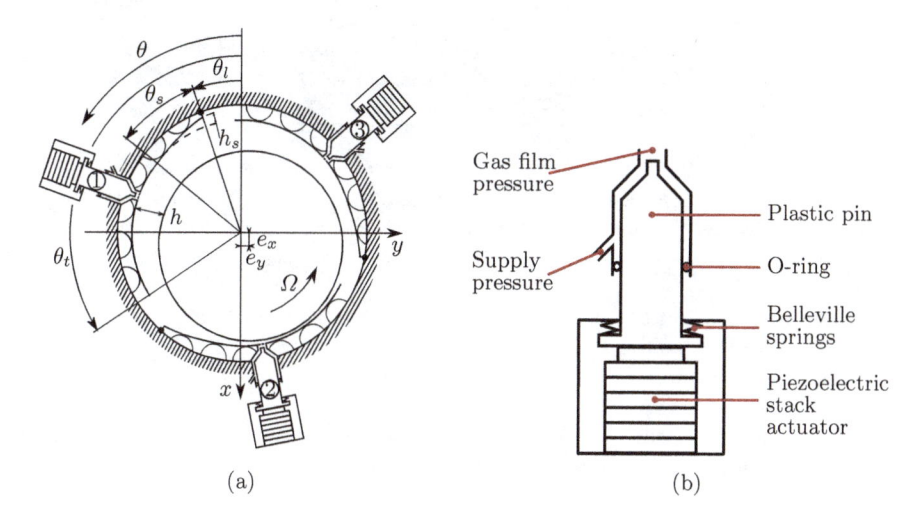

Fig. 2. (a) Sketch of the hybrid gas foil bearing geometry with relevant geometrical parameters defined. The injectors are placed at the pad centers at 87.5° for injector 1, 207.5° for injector 2, and 327.5° for injector 3. Recreated from Heinemann et al. [24]. (b) The piezoelectric injection actuator components. Recreated from von Osmanski and Santos [25].

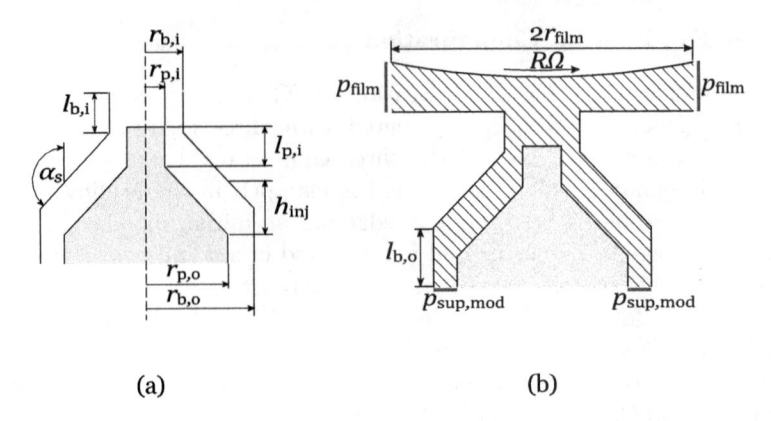

Fig. 3. (a) Geometrical dimensions of the rotationally symmetrical Computational Fluid Dynamics (CFD) model of the injector. (b) Sketch of the cross-sectional area of the injector CFD model (hatched area), including the boundary conditions, where the fluid film breaks the rotational symmetry. Both recreated from von Osmanski and Santos [25].

Table 1. Geometry, material properties, and operating condition of the GFB test rig, illustrated in Fig. 1 and 2a.

Shaft assembly			
Bearing A to CG, l_1	201.1 mm	Mass, m_r	21.17 kg
Bearing B to CG, l_2	197.9 mm	Polar moment of inertia, I_{zz}	0.030 08 kg m^2
Disc A to CG, l_3	287.2 mm	Transverse moment of inertia,	0.5252 kg m^2
Disc B to CG, l_4	304.0 mm	$I_{xx} = I_{yy}$	
Bearing configuration			
Bearing radius, R	33.50 mm	Slope extent, θ_s	30°
Bearing length, L	53.00 mm	Slope height, h_s	50 µm
Radial clearance, C	40 µm	First pad leading edge, θ_l	30°
Number of pads, n_{pad}	3	First pad trailing edge, θ_t	115° + θ_l
Fluid properties			
Viscosity, μ	1.95 × 10^{-5} Pa s	Specific gas constant, R_s	287.1 J kg^{-1} K^{-1}
Ambient pressure, p_a	10^5 Pa		
Foil structure properties			
SEFM stiffness, k	8.8 GN m^{-3}	Loss factor, η	0.15
Top foil thickness, t_t	0.254 mm	Foil oscillation frequency, ω_f	Ω
Young's modulus, E_t	207.0 GPa	Bump foil pitch	7.0 mm

3 Mathematical Modelling

The mathematical modeling of the system, including the foil, rotor, and fluid domains is described in von Osmanski and Santos [28], with injection incorporated in von Osmanski and Santos [25], and actuator states by Heinemann [26]. The modeling is briefly described here for the reader's convenience.

Table 2. Geometry, material properties, and operating condition of the proposed GFB Injectors illustrated in Fig. 3.

Injector geometry			
Nozzle clearance, h_{inj}	$0 - 45\,\mu\text{m}$	Radius of film zone, r_{film}	2.5 mm
Inner bore radius, $r_{\text{b,i}}$	1.0 mm	Outer bore radius, $r_{\text{b,o}}$	3.0 mm
Inner pin radius, $r_{\text{p,i}}$	0.85 mm	Outer pin radius, $r_{\text{p,o}}$	2.7 mm
Bore length at inner radius, $l_{\text{b,i}}$	1.37 mm	Bore length at inner radius, $l_{\text{b,o}}$	2.0 mm
Rounding radius, bore-film edge	$2.0\,\mu\text{m}$	Rounding radius, other edges	$10\,\mu\text{m}$
Pin length at inner radius, $l_{\text{p,i}}$	1.2 mm	Shoulder angle, α_s	$120°$
Actuator properties			
Belleville washers stiffness, k_{mech}	5×10^6 N/m	Piezoelectric element stiffness, k_{piezo}	3×10^7 N/m
Belleville washers and O-ring equivalent damping, d_{mech}	2×10^3 N/ms	Piezoelectric coupling coefficient, c_{piezo}	65×10^{-8} C/N
Pin mass, m_{pin}	0.05 kg	No. of piezoelectric elements in stack, N_{stack}	100

3.1 Multi Domains

Fluid Film. The fluid film is modeled using the compressible and transient Modified Reynolds Equation (MRE) for an ideal gas. The air is assumed isothermal and isoviscous, and the density is described through the ideal gas law. From the non-dimensionalized MRE, the Finite Volume (FV) residual formulation is derived, with the Control Volume Center pressures (CVC) as states. The pressure is non-dimensionalized with the ambient pressure p_a, and time with the characteristic frequency $\omega_\tau = \Omega$. The FV residual equation for bearing α is;

$$\mathbf{A}_{\text{FV}}\left(\tilde{p}_\alpha, \tilde{x}_r, \tilde{x}_{f,\alpha}\right) \tilde{p}_\alpha + \mathbf{B}_{\text{FV}}\left(\dot{\tilde{x}}_r, \dot{\tilde{x}}_{f,\alpha}\right) \tilde{p}_\alpha \\ + \mathbf{C}_{\text{FV}}\left(\tilde{x}_r, \tilde{x}_{f,\alpha}\right) \dot{\tilde{p}}_\alpha + \dot{\tilde{m}}_\alpha = \tilde{r}_{\text{FV},\alpha} \quad \in \mathbb{R}^{n_{\text{CV}}}, \qquad (1)$$

where $\mathbf{A}_{\text{FV}}$ represents the Couette and Poiseuille coefficients, and $\mathbf{B}_{\text{FV}}$ and $\mathbf{C}_{\text{FV}}$ represent the local expansion and squeeze coefficients respectively of the MRE. $\dot{\tilde{m}}_\alpha$ describes the mass flows from the injectors. The axial and tangential fluid film components are evaluated at the midpoint of the Control Volume Edges (CVEs). The face midpoint pressure gradients are reconstructed from the CVC pressures using a Central Differencing Scheme (CDS), whereas the face midpoint pressure is determined using a Linear Upwind Differencing Scheme (LUDS). A non-uniform structured Cartesian grid is used to capture the steep pressure gradients at the injection holes.

Injection. The model of the injector is shown in Fig. 3b and relies on the assumptions of; 1) laminar and steady flow, 2) the pressure along the circumference of the injection zone with radius r_{film} is uniform, and 3) the film height within the sub-model is constant. The wall boundary temperature is assumed constant and the model is solved with a compressible solver.

Solutions for a parameter space were created by von Osmanski and Santos [25], spanning several values of five, dominant parameters; 1) the nozzle clearance, h_{inj}, 2) modified supply pressure $p_{sup,mod}$, 3) fluid film pressure p_{film}, 4) eccentricity ϵ_x, and 5) rotor angular velocity Ω. The parameter space values can be seen in Table 3. The interpolated injection mass flow through the injectors can thereby be described through a linear interpolation function of the parameter space. The derivatives of the injection flow with respect to the dominant parameters are likewise determined through interpolation. To determine the sub-model parameters, constant transformation matrices and scaling vectors are used for the fluid film pressure and bearing eccentricity to determine the sub-model parameters, whereas the injector valve clearance, modified supply pressure, and rotating velocity are directly available from the rotor-bearing model.

Table 3. The precalculated 5-dimensional parameter space.

Parameter	Levels	Values
Nozzle clearance, h_{inj} [μm]	7	5.0, 10.0, 15.0, 20.0, 25.0, 30.0, 35.0
Mod. supply pressure, $p_{sup,mod}$ [kPa]	4	150, 233, 317, 400
Fluid film pressure, p_{film} [kPa]	5	90.0, 168, 245, 323, 400
Eccentricity, ϵ_x [−]	7	−0.9, −0.5, 0.0, 0.5, 0.7, 0.8, 0.9
Angular velocity, Ω [kRPM]	3	0, 20, 40

Rotor. The rotor is modeled as a rigid shaft with four Degrees of Freedom (DOFs), the horizontal y, and vertical x displacement in either bearing, where the DOFs are non-dimensionalized with the bearing clearance. Given that the compliant rotor dynamics are considerably higher than the maximum rotational velocity [23], minimal loss of accuracy is achieved with a rigid shaft assumption for the frequency domain of interest. The rotor residual equation, written in state space residual form becomes,

$$\begin{bmatrix} \mathbf{I} & \mathbf{0} \\ \mathbf{0} & \tilde{\mathbf{M}}_r \end{bmatrix} \begin{Bmatrix} \dot{\tilde{x}}_r \\ \ddot{\tilde{x}}_r \end{Bmatrix} - \begin{bmatrix} \mathbf{0} & \mathbf{I} \\ \mathbf{0} & \Omega\tilde{\mathbf{G}}_r \end{bmatrix} \begin{Bmatrix} \tilde{x}_r \\ \dot{\tilde{x}}_r \end{Bmatrix} - \begin{Bmatrix} \mathbf{0} \\ \tilde{\boldsymbol{f}}_{rw} + \tilde{\boldsymbol{f}}_b + \tilde{\boldsymbol{f}}_{ub} \end{Bmatrix} = \tilde{\boldsymbol{r}}_r \quad \in \mathbb{R}^8, \quad (2)$$

where $\tilde{\boldsymbol{r}}_r$ is the residual, and $\tilde{\mathbf{M}}_r$ and $\tilde{\mathbf{G}}_r$ are the non-dimensionalized rotor mass and gyroscopic matrices respectively, while $\tilde{\boldsymbol{f}}_{rw}$, $\tilde{\boldsymbol{f}}_b$, and $\tilde{\boldsymbol{f}}_{ub}$ are static, bearing, and unbalance forces respectively. The bearing forces are obtained through the integration of the projected air film relative pressure. The integration of the bearing forces can be described as a non-constant mapping by the enforcement of the Gümbel condition. The rigid film height contribution at both the CVCs and CVE midpoints is required in the FV Eq. (1). The rotor movement in the bearing is assumed to be axially uniform, causing the film height to solely depend on the circumferential coordinate. Therefore, the projections from the rotor position to the resulting rigid film height can be described through a trigonometric function. Given the fluid film FV residual Eq. (1), the projection from the rotor position to the CVC is required for the squeeze and local expansion terms, and CVE for the Couette and Poiseuille terms, can be described through the constant mappings shown in Osmanski et al. [23].

Foil Structure. The top foil is modeled using massless 2D Euler-Bernoulli (EB) beam elements solely with rotational and transverse DOFs, given the good approximation of the stiffening effects of curved shells, as shown by Heinemann et al. [24]. The bump foil support is modeled using the Simple Elastic Foundation Model (SEFM), described by Heshmat and Walowit [29, 30], with constant stiffness per unit area k and constant loss factor η in both the axial and angular directions of the pad. The stiffness and damping of the SEFM are added to the EB beams through work equivalent nodal forces [25].

The foil domain in each bearing is discretized into n_fe foil elements spanning the circumferential length of the domain in the bearing. With the leading edge clamped the foil DOF vector x_f size, n_fdof, becomes twice the number of elements in the domain as each node contributes two DOF. The global foil structure stiffness $\tilde{\mathbf{K}}_f$, and damping $\tilde{\mathbf{D}}_f$ matrices are constructed as,

$$\tilde{\mathbf{K}}_f = \sum_e \left(\tilde{\mathbf{k}}_t^e + \tilde{\mathbf{k}}_\text{SEFM}^e \right), \quad \tilde{\mathbf{D}}_f = \eta \tilde{\omega}_f^{-1} \sum_e \tilde{\mathbf{k}}_\text{SEFM}^e, \qquad (3)$$

where $\tilde{\mathbf{k}}_t^e$ is the top foil element stiffness matrix, and $\tilde{\mathbf{k}}_\text{SEFM}^e$ the SEFM equivalent element stiffness matrix. The element matrices are non-dimensionalized using the ambient pressure p_a, element length l_e, and time scale ω_τ^{-1}. The damping is introduced using the equivalent viscous damping based on the dissipated energy per cycle, [31], where it is assumed that the foil oscillates at the frequency of the rotation velocity i.e. $\omega_f = \Omega$. The residual foil equation of motion for bearing α can be written as,

$$\tilde{\mathbf{D}}_f \dot{\tilde{\boldsymbol{x}}}_{f,\alpha} + \tilde{\mathbf{K}}_f \tilde{\boldsymbol{x}}_{f,\alpha} - \tilde{\boldsymbol{f}}_{p,\alpha} = \tilde{\boldsymbol{r}}_{f,\alpha}, \quad \in \mathbb{R}^{n_\text{fdof},\alpha}. \qquad (4)$$

The full foil load vector in the bearing α from the bearing pressure distribution can be determined by using a constant mapping matrix as described by von Osmanski and Santos [25], from the work equivalent nodal forces. Using the element shape functions, the film height contributions at the CVC and CVE are similarly determined with constant linear mappings as described by von Osmanski and Santos [25].

Actuator. The piezoelectric actuator is modeled as described by Morosi and Santos [32] using the linear constitutive law for piezoelectric materials to describe the piezoelectric force as described in the ANSI-IEEE 176 Standard of Piezoelectricity [33]. The pin dynamics are modeled as a second-order mass-spring-damper system.

The pin dynamics for each injector can be described in state space format as [26],

$$\begin{bmatrix} \ddot{x}_{\text{inj},i} \\ \dot{x}_{\text{inj},i} \end{bmatrix} - \begin{bmatrix} 0 & 1 \\ -\frac{k_{\text{piezo}}+k_{\text{mech}}}{m_{\text{pin}}} & -\frac{d_{\text{mech}}}{m_{\text{pin}}} \end{bmatrix} \begin{bmatrix} \dot{x}_{\text{inj},i} \\ x_{\text{inj},i} \end{bmatrix} - \begin{bmatrix} 0 \\ \frac{k_{\text{piezo}} c_{\text{piezo}} N_{\text{stack}}}{m_{\text{pin}}} \end{bmatrix} u_{\text{inj},i} = r_{\text{inj},i} \in \mathbb{R}^2. \tag{5}$$

The injector residual is non-dimensionalized using the system timescale ω_τ^{-1}, and with the maximum injector valve clearance, $h_{\text{inj,max}}$, as the actuator DOF lengthscale. The bearing actuator states for bearing α are collected to determine the bearing actuator residual $\tilde{x}_{\text{inj},\alpha}$ for bearings α. The α bearing actuator state vector is;

$$\tilde{x}_{\text{inj},\alpha} = h_{\text{inj,max}}^{-1} \{x_{\text{inj},1}^T \ x_{\text{inj},2}^T \ \cdots \ x_{\text{inj},n_{\text{inj}}}^T\}^T. \tag{6}$$

The relation between the injector valve clearance h_{inj} and the actuator position can be described as, $h_{\text{inj}} = h_{\text{inj,max}} - x_{\text{inj},i}(t)$, as seen in Fig. 3.

3.2 Full Coupled System

The model domains are collected in a global state system of size n_{DOF} in order to solve the full system. The resulting state vector becomes,

$$z = \{\tilde{p}_A^T \ \tilde{p}_B^T \ \tilde{x}_{f,A}^T \ \tilde{x}_{f,B}^T \ \tilde{x}_r^T \ \dot{\tilde{x}}_r^T \ \tilde{x}_{\text{inj},A}^T \ \tilde{x}_{\text{inj},B}^T\}^T \in \mathbb{R}^{n_{\text{DOF}}}, \tag{7}$$

with the residual vector,

$$\tilde{r} = \{\tilde{r}_{\text{FV},A}^T \ \tilde{r}_{\text{FV},B}^T \ \tilde{r}_{f,A}^T \ \tilde{r}_{f,B}^T \ \tilde{r}_r^T \ \tilde{r}_{\text{inj},A}^T \ \tilde{r}_{\text{inj},B}^T\}^T \in \mathbb{R}^{n_{\text{DOF}}}. \tag{8}$$

Defining the rotor external rotor unbalance as $v = \tilde{f}_{\text{ub}} + \tilde{f}_{\text{rw}}$, for later convenience. The collected contribution from the domains gives rise to the system residual description,

$$f_G(z, \dot{z}, u, v) = f_{G,0} + f_{G,s}(z) + f_{G,t}(z, \dot{z}) + f_{G,u}(u) + f_{G,v}(v) = \tilde{r}. \tag{9}$$

where the residual can be split into its constant part $f_{G,0}$, its steady part $f_{G,s}(z)$, its transient part $f_{G,t}(z,\dot{z})$, its input dependent part $f_{G,u}(u)$, and the part dependant on the external disturbance $f_{G,v}(v)$. The system Jacobian can correspondingly be partitioned in steady-state Jacobian $\partial f_{G,s}/\partial z = J_{z,s}$ and the time-dependant Jacobians with respect to z and $\dot{z}$, i.e. $\partial f_{G,t}/\partial z = J_{z,t}$ and $\partial f_{G,t}/\partial \dot{z} = J_{\dot{z}}$ respectively. The Jacobian with respect to the state is collectively defined as the summation of the steady Jacobian $J_{z,s}$ and the transient Jacobian $J_{z,t}$, i.e. $J_z = J_{z,s} + J_{z,t}$. The Jacobian with respect to the input force and external disturbance are similarly defined as, $\partial f_{G,u}/\partial u = J_{\dot{u}}$ and $\partial f_{G,v}/\partial \dot{v} = J_{\dot{v}}$ respectively. A detailed description of the derivatives is given in von Osmanski and Santos [25].

3.3 Linearization

To create a controller for the non-linear system, the system is linearized and the system matrix $\mathbf{A}$, the excitation matrix $\mathbf{B}$ and disturbance matrix $\mathbf{B}_v$ are determined. Taylor expanding around the static equilibrium position, $\dot{z}_0 = \mathbf{0}$ results in the linearized system description,

$$\boldsymbol{f}_G(\boldsymbol{z}, \dot{\boldsymbol{z}}, \boldsymbol{u}, \boldsymbol{v}) \approx \mathbf{J}_{\dot{z}}\Delta\dot{\boldsymbol{z}} + \mathbf{J}_z\Delta\boldsymbol{z} + \mathbf{J}_u\Delta\boldsymbol{u} + \mathbf{J}_v\Delta\boldsymbol{v}. \tag{10}$$

Assuming the system residuals to be zero after convergence and isolating hereafter for the highest order derivative, the matrices $\mathbf{A}$, $\mathbf{B}$ and $\mathbf{B}_v$ are readily available,

$$\Delta\dot{\boldsymbol{z}} = \underbrace{(-\mathbf{J}_{\dot{z}}^{-1}\mathbf{J}_z)}_{\mathbf{A}}\Delta\boldsymbol{z} + \underbrace{(-\mathbf{J}_{\dot{z}}^{-1}\mathbf{J}_u)}_{\mathbf{B}}\Delta\boldsymbol{u} + \underbrace{(-\mathbf{J}_{\dot{z}}^{-1}\mathbf{J}_v)}_{\mathbf{B}_v}\Delta\boldsymbol{v}. \tag{11}$$

The linear stability and linear system characteristics at the linearization point $\boldsymbol{z}_0$ can be determined by solving the standard eigenvalue problem $-\mathbf{J}_{\dot{z}}^{-1}\mathbf{J}_z \boldsymbol{t}_i = \lambda_i \boldsymbol{t}_i$.

Based on the resulting eigenvalues λ_i and eigenvectors $\boldsymbol{t}_i$ the linear system dynamics can be assessed. Given that the rotor position DOF of the system is observed in both the x- and y - direction, i.e. $\boldsymbol{x}_r$, the corresponding output matrix $\mathbf{C}$, therefore becomes:

$$\mathbf{C} = \begin{bmatrix} 0 & 0 & 0 & 0 & \mathbf{I} & 0 & 0 & 0 \end{bmatrix}. \tag{12}$$

The linearized system can be described through,

$$\begin{cases} \Delta\dot{\boldsymbol{z}} = \mathbf{A}\Delta\boldsymbol{z} + \mathbf{B}\Delta\boldsymbol{u} + \mathbf{B}_v\Delta\boldsymbol{v}, \\ \Delta\boldsymbol{y} = \mathbf{C}\Delta\boldsymbol{z}. \end{cases} \tag{13}$$

3.4 Model Reduction

In order to create a system controller that is applicable and adequately fast for physical system implementation, a model reduction is performed, projecting the system onto the four complex conjugated rotor modes, through modal subspace projection, as done by Christensen and Santos [34, 35]. The collection of the real and imaginary part of the $q = 4$ corresponding right $\boldsymbol{t}$ and left $\boldsymbol{l}$ eigenvectors result in the projection matrices,

$$\mathbf{V}_R = \begin{bmatrix} \mathrm{Re}(\boldsymbol{t}_1) & \mathrm{Im}(\boldsymbol{t}_1) \ldots \mathrm{Re}(\boldsymbol{t}_q) & \mathrm{Im}(\boldsymbol{t}_q) \end{bmatrix}$$
$$\text{and} \tag{14}$$
$$\mathbf{W}_R = \begin{bmatrix} \mathrm{Re}(\boldsymbol{l}_1)^\mathrm{T} & \mathrm{Im}(\boldsymbol{l}_1)^\mathrm{T} \cdots \mathrm{Re}(\boldsymbol{l}_q)^\mathrm{T} & \mathrm{Im}(\boldsymbol{l}_q)^\mathrm{T} \end{bmatrix}^\mathrm{T}.$$

The system described in the rotor approximated coordinates thereby becomes,

$$\Delta\dot{\boldsymbol{z}}_R(t) = \mathbf{A}_R\Delta\boldsymbol{z}_R(t) + \mathbf{B}_R\Delta\boldsymbol{u}(t) + \mathbf{B}_{v,R}\boldsymbol{v}(t) , \quad \Delta\boldsymbol{y}_R(t) = \mathbf{C}_R\Delta\boldsymbol{z}_R(t), \tag{15}$$

where the projection between the full and reduced system states become, assuming that both rotor position and velocity states are observed, i.e.
$\mathbf{C}_{\text{red}} = [0\ 0\ 0\ 0\ \mathbf{I}\ \mathbf{I}\ 0\ 0]$,

$$\begin{aligned}
\mathbf{A}_R &= \mathbf{C}_{\text{red}} \mathbf{V}_R \left(\mathbf{W}_R^T \mathbf{V}_R\right)^{-1} \mathbf{W}_R^T \mathbf{A} \mathbf{V}_R \left(\mathbf{C}_{\text{red}} \mathbf{V}_R\right)^{-1}, \\
\mathbf{B}_R &= \mathbf{C}_{\text{red}} \mathbf{V}_R \left(\mathbf{W}_R^T \mathbf{V}_R\right)^{-1} \mathbf{W}_R^T \mathbf{B}, \\
\mathbf{B}_{v,R} &= \mathbf{C}_{\text{red}} \mathbf{V}_R \left(\mathbf{W}_R^T \mathbf{V}_R\right)^{-1} \mathbf{W}_R^T \mathbf{B}_v, \quad \mathbf{C}_R = \mathbf{I}.
\end{aligned} \quad (16)$$

3.5 Control Architecture

To implement classical linear control theory, the injector stationary equilibrium position $z_{inj,0}$, with the injector position $h_{inj} = \frac{1}{2} h_{inj,\max}$, i.e. the injectors half open, is chosen as the control position. This allows the injectors to move equally in both directions.

The controller is designed in the unstable region at 35 kRPM to ensure the controller's intended dynamics at the important region. The controller will, thereby, be invariant with the rotational velocity. From the current setup, only rotor position data is collected, thus only Δx_r is directly available for the controller. The designed controller and required observer for $\Delta \dot{x}_r$ is built based on the reduced linear system model. The observer and controller are designed separately, made possible by the separation principle [36]. As the presented study is a feasibility study a deterministic continuous-time controller and observer architecture is designed.

Control Design. The desired effect of the control architecture is increased damping and a retarding of the Onset Speed of Instability (OSI). The control architecture is therefore based on linear full-state feedback of the reduced state. The reduced order state vector, $\Delta z_R \approx \{\Delta x_r^T, \Delta \dot{x}_r^T\}^T$, is thereby required to be available for the control scheme. The control input is defined as,

$$\Delta u = -\mathbf{K} \Delta z_R, \quad (17)$$

where $\mathbf{K}$ in the controller gain matrix.

The system gain matrix is readily determined using steady-state linear Quadratic Regulator (LQR) theory [36], penalizing the rotor forwards modes, as the reduced system is controllable.

Observer Design. As the current system is only fitted with rotor position sensors, the use of full-state feedback would require a rotor velocity state estimator. The reduced system is partitioned into the rotor position approximation Δz_1 and rotor velocity approximation Δz_2 for later convenience,

$$\begin{cases} \begin{bmatrix} \Delta \dot{z}_1 \\ \Delta \dot{z}_2 \end{bmatrix} = \begin{bmatrix} \mathbf{A}_{11} & \mathbf{A}_{12} \\ \mathbf{A}_{21} & \mathbf{A}_{22} \end{bmatrix} \begin{bmatrix} \Delta z_1 \\ \Delta z_2 \end{bmatrix} + \begin{bmatrix} \mathbf{B}_1 \\ \mathbf{B}_2 \end{bmatrix} \Delta u, \\ \Delta y_R = \mathbf{C}_1 \Delta z_1, \end{cases} \quad (18)$$

where $\mathbf{C}_1 = \mathbf{I}$, as only rotor position is observed. A reduced order observer can be readily built for Δz_2, as the reduced system is fully observable. The state estimator, $\Delta \hat{z}_2$ can be described as,

$$\Delta \hat{z}_2 = z_{\text{aux}} + \mathbf{L} \Delta y_R, \tag{19}$$

where $\mathbf{L}$ is the observer gain matrix and the auxiliary state vector z_{aux} is defined as,

$$\dot{z}_{\text{aux}} = \mathbf{M} z_{\text{aux}} + \mathbf{N} \Delta u + \mathbf{P} \Delta y_R. \tag{20}$$

The constant matrices $\mathbf{M}$, $\mathbf{N}$, and $\mathbf{P}$, are defined as [36],

$$\begin{aligned}
\mathbf{M} &= \mathbf{A}_{22} - \mathbf{L}\mathbf{C}_1\mathbf{A}_{12}, \\
\mathbf{N} &= \mathbf{B}_2 - \mathbf{L}\mathbf{C}_1\mathbf{B}_1, \\
\mathbf{P} &= (\mathbf{A}_{21} - \mathbf{L}\mathbf{C}_1\mathbf{A}_{11})\mathbf{C}_1^{-1} + \mathbf{M}\mathbf{L}.
\end{aligned} \tag{21}$$

The observer design reduces to the determination of $\mathbf{L}$ such that the system observer has the desired behavior. For a fast dynamic, the imaginary part of the observer eigenvalues is placed 10 times higher than the closed loop counterpart. Requiring that the damping ratios are $\zeta_o = \sqrt{2}/2$ a well-behaved observer is obtained [36]. The observer gain matrix required to obtain the eigenvalues is determined using Eigenstructure Assignment [36].

3.6 Combined Observer and Controller on the Full Non-linear System

In order to append the observer and controller to the full non-linear system, the two are first combined in a linear model. Given the auxiliary state Eq. (20) for the observer, the combined system control input, with observer and controller, can be described as,

$$\Delta u = -\begin{bmatrix} \mathbf{K}_1 & \mathbf{K}_2 \end{bmatrix} \begin{bmatrix} \Delta z_1 \\ \Delta \hat{z}_2 \end{bmatrix} = -(\mathbf{K}_1 + \mathbf{K}_2 \mathbf{L} \mathbf{C}_1)\Delta z_1 - \mathbf{K}_2 z_{\text{aux}}. \tag{22}$$

In order to incorporate the observer into the non-linear system, the auxiliary state equation is written in residual form,

$$\dot{z}_{\text{aux}} - (\mathbf{M} - \mathbf{N}\mathbf{K}_2)z_{\text{aux}} - (\mathbf{P}\mathbf{C}_1 - \mathbf{N}(\mathbf{K}_1 + \mathbf{K}_2\mathbf{L}\mathbf{C}_1))(\tilde{x}_r - \tilde{x}_{r,0}) = r_{\text{aux}} \quad \in \mathbb{R}^4. \tag{23}$$

Adding the observer equation to the full coupled system the resulting closed-loop state vector becomes,

$$z_{\text{cl}} = \left\{ \tilde{p}_A^T \; \tilde{p}_B^T \; \tilde{x}_{f,A}^T \; \tilde{x}_{f,B}^T \; \tilde{x}_r^T \; \dot{\tilde{x}}_r^T \; \tilde{x}_{\text{inj},A}^T \; \tilde{x}_{\text{inj},B}^T \; z_{\text{aux}}^T \right\}^T, \tag{24}$$

and the corresponding residual vector is,

$$\tilde{r}_{\text{cl}} = \left\{ \tilde{r}_{\text{FV},A}^T \; \tilde{r}_{\text{FV},B}^T \; \tilde{r}_{f,A}^T \; \tilde{r}_{f,B}^T \; \tilde{r}_r^T \; \tilde{r}_{\text{inj},A}^T \; \tilde{r}_{\text{inj},B}^T \; r_{\text{aux}}^T \right\}^T. \tag{25}$$

The full system augmented with the control architecture becomes,

$$f_{G,cl}(z_{cl}, \dot{z}_{cl}, v) = f_{G,t,cl}(\tau, z, \dot{z}) + f_{G,0,cl} + f_{G,v,cl} \\ + f_{G,s,cl}(z) = \tilde{r}_{cl}. \quad (26)$$

Similarly to the open-loop system, the corresponding Jacobians for the controlled system, $\mathbf{J}_{cl}$, can be defined, and used to describe the closed-loop linearized system,

$$\Delta \dot{z}_{cl} = \underbrace{(-\mathbf{J}_{\dot{z},cl}^{-1}\mathbf{J}_{z,cl})}_{\mathbf{A}_{cl}}\Delta z_{cl} + \underbrace{(-\mathbf{J}_{\dot{z},cl}^{-1}\mathbf{J}_{v,cl})}_{\mathbf{B}_{v,cl}}\Delta v. \quad (27)$$

The closed loop linear stability and linear system characteristics at the linearization point z_0 can be determined by solving the standard eigenvalue problem $-\mathbf{J}_{\dot{z},cl}^{-1}\mathbf{J}_{z,cl}t_i = \lambda_i t_i$.

4 Theoretical Results

Mesh convergence was assessed by von Osmanski and Santos [25], which showed that the OSI converged slower than other system properties such as the equilibrium position with respect to the fluid film and top foil meshes. Convergence of the OSI was found to require five CVs across the injection zone and a reasonable grading, resulting in a mesh with $n_z = 18$ and $n_\theta = 135$ per pad. The converged mesh can be seen for the first pad in Fig. 4. Cook [37] describes that shell elements should not span more than 15°, extending this criterion to beam elements results in a minimum of 8 elements per pad. Convergence was, however, already archived with 5 beam elements per pad. Given the low cost of additional foil elements, 10 elements are used per pad giving $n_{fe} = 30$. The full system is comprised of 2430 CVs (fluid DOFs) and 60 foil DOFs per bearing, 8 rotor states, and 6 actuator states per bearing resulting in a full system of 5000 states. Three test cases are here presented, (a) the passive GFB, (b) the passive HGFB with injector 3 fully open, and (c) the active HGFB. The system is modeled for the constant injection pressure of $p_{inj} = 300$ kPa. The equilibrium position is determined using a Newton-Raphson method with a trust-region scheme [38] through the Matlab function `fsolve`. The transient solution is determined by using a variable-step, variable-order, 1–5, method based on the backward differencing formula (BDF) [39], using the Matlab implicit solver `ode15i`, and the equilibrium position as the initial value. Both the equilibrium position and the transient solution methods make use of the analytically defined system Jacobians.

4.1 Non-linear Steady-State Unbalance Response

The assessment of the active system is obtained from the major axis unbalance response of the three systems. The systems are exited with the unbalance, $ub = \{-2.5\ 14\}$gmm, corresponding to an ISO rating of G6.3. Large

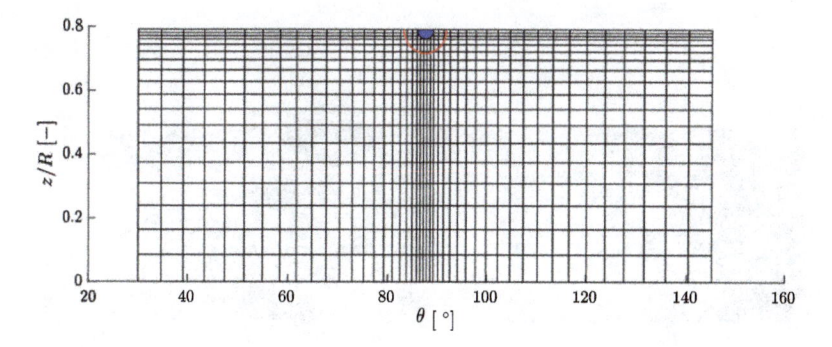

Fig. 4. Fluid film (FV) discretization, shown for bearing pad 1, with the filled blue (●) semicircle being the injection zone where, $q_{\text{inj}} \neq 0$ and $\dot{m}_i = \int \int_{A_i} \tilde{q}_{\text{inj}} \mathrm{d}A$ is added as a source term to the CV residuals. The red line (——) describes the CFD sub-model boundary radius, thus being the boundary at which the pressure is assumed uniform along the circumference. (Color figure online)

sub-synchronous unbalance amplitudes are seen in the passive system response at $\Omega \approx 13\,\text{kRPM}$, as seen in Fig. 5a. The use of the third injector shows that the sub-synchronous vibrations can be completely suppressed at this unbalance level, as seen in Fig. 5b. The maximum major axis is, however, increased from $5.64\,\mu\text{m}$ to $6.23\,\mu\text{m}$, as stated in Table 4. In both the passive and the hybrid passive systems, transient vibrations in the higher rotational velocities can be observed, as the damping ratios decrease towards the OSI. The reduced damping causes the transient response to spill into the assumed steady response. Assessing the unbalance response waterfall diagram of the controlled system, Fig. 5c, the sub-synchronous vibrations are likewise suppressed. Moreover, the maximum major axis is markedly lower than those of the passive GFB and HGFB, being $4.11\,\mu\text{m}$.

Table 4. Maximum major axis amplitudes of the three cases in the operational range $\Omega \in [5,\ 29]\,\text{kRPM}$, and frequency ranges $[0,\ 500]\,\text{Hz}$ for both bearing A and B.

System	B [μm]	A [μm]
No Inj.	4.74	5.64
Inj. 3 open	5.64	6.23
Controlled	2.83	4.11

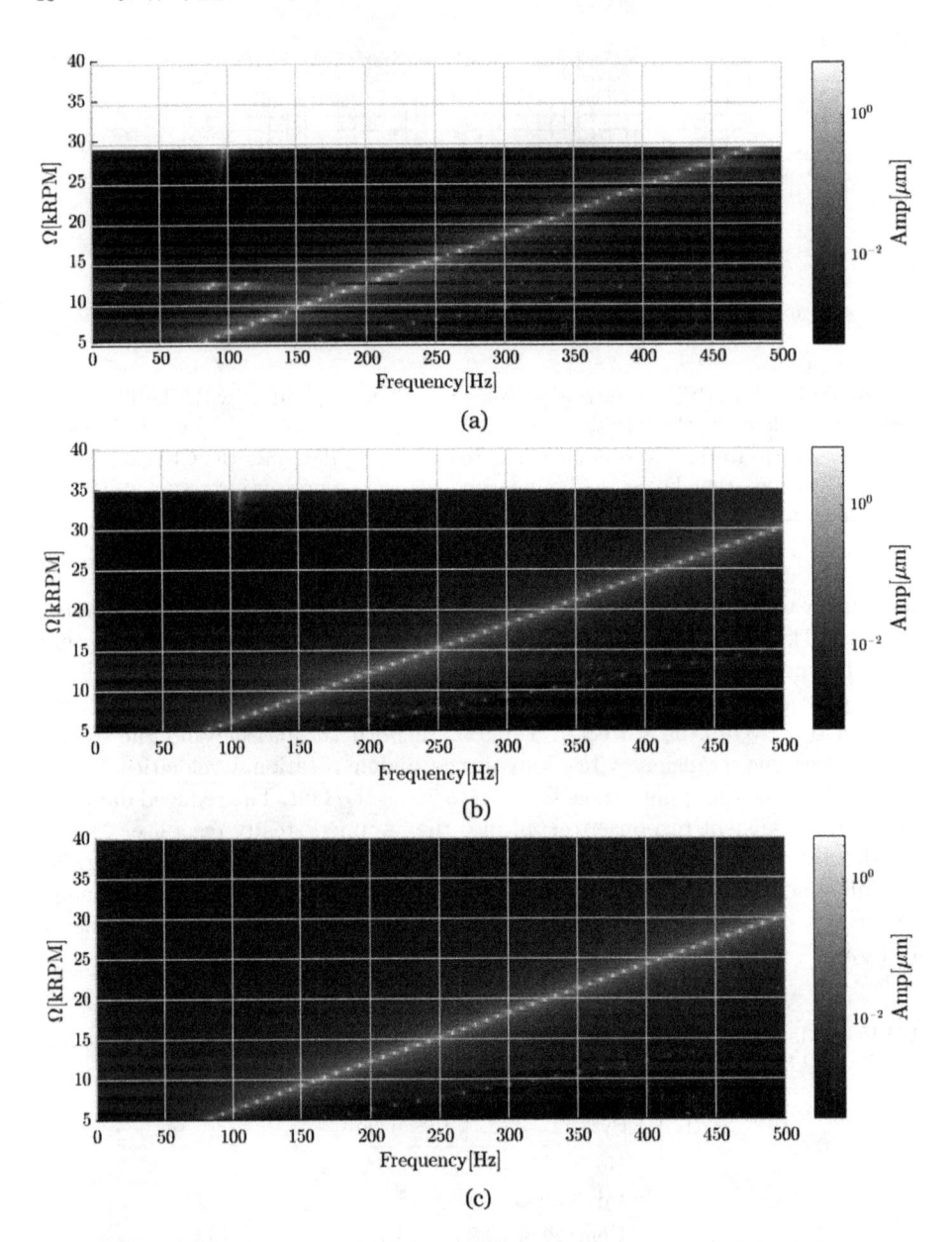

Fig. 5. The major axis amplitude bearing A unbalance response waterfall diagram of the (a) passive system with no injection in the operating range $\Omega \in [5, 29.5]$kRPM, (b) the hybrid passive system in the operating range $\Omega \in [5, 35]$kRPM, and (c) the actively controlled system in the operating range $\Omega \in [5, 40]$kRPM. For the passive system, the state is approximated by the time response from $1.0 - 3.125$ s, whereas for the hybrid and active systems the steady state is approximated by the time response from $0.5 - 2.254$ s. The systems are excited with unbalance $\boldsymbol{ub} = \{-2.5 \ 14\}$gmm, corresponding to a G6.3 unbalance rating defined at $\Omega = 40$ kRPM.

5 Conclusion

The main contribution of this paper is to present a model-based controller for a system using active HGFBs with radial gas injection. The control system architecture was designed based on the linearization and model reduction of the current non-linear mathematical model, and implementation of classical control theory. The effects of utilizing a passive hybrid system for a specific case were analyzed, through an unbalance response waterfall diagram, showing that the use of the third injector could increase the system stability for increased operational velocities. The controlled system with the designed controller resulted in a stable system in the assessed operational range of interest, i.e. $\Omega \in [5, 40]$kRPM capable of suppressing sub-synchronous vibration at the ISO unbalance rating of G6.3, and significantly reducing the steady-state unbalance vibrations by 27% in comparison to the passive GFB system, and 34% in comparison to the passive HGFB system, with respect to vibrations in bearing A, in range $\Omega \in [5, 29]$kRPM.

Further work on the current project is focused on assessing the load-bearing capacities of the design configurations, in addition to the effect the configurations have on the modal parameters. Parallel experimental verification of the numerical model, with the inclusion of the injectors for the HGFB, is ongoing, as the GFB and hybrid gas-bearing numerical models are already experimentally verified [22, 25, 40].

References

1. Oh, K.P., Rohde, S.M.: A theoretical investigation of the multileaf journal bearing. J. Appl. Mech. **43**, 237–242 (1976)
2. Heshmat, H., Walowit, J.A., Pinkus, O.: Analysis of gas-lubricated foil journal bearings. J. Lubr. Technol. **105**, 647–655 (1983)
3. Peng, J., Carpino, M.: Calculation of stiffness and damping coefficients for elastically supported gas foil bearings. J. Tribol. **115**, 20–27 (1993)
4. Schiffmann, J., Spakovszky, Z.S.: Foil bearing design guidelines for improved stability. J. Tribol. **135**, 12 (2012)
5. Pattnayak, M., Ganai, P., Pandey, R., Dutt, J., Fillon, M.: An overview and assessment on aerodynamic journal bearings with important findings and scope for explorations. Tribol. Int. **174**, 107778 (2022)
6. Adolfo, D., Ertas, B.: Dynamic characterization of a novel externally pressurized compliantly damped gas-lubricated bearing with hermetically sealed squeeze film damper modules. In: ASME Turbo Expo 2018: Turbomachinery Technical Conference and Exposition, p. V07BT34A050 (2018)
7. Ertas, B., Adolfo, D.: Compliant hybrid gas bearing using modular hermetically sealed squeeze film dampers. J. Eng. Gas Turbines Power **141**, 08 (2018)
8. Feng, K., Cao, Y., Yu, K., Hanqing, G., Wu, Y., Guo, Z.: Characterization of a controllable stiffness foil bearing with shape memory alloy springs. Tribol. Int. **136**, 03 (2019)
9. Yazdi, B.Z., Kim, D.: Rotordynamic performance of hybrid air foil bearings with regulated hydrostatic injection. In: Journal of Engineering for Gas Turbines and Power-transactions of the ASME, p. V07AT34A012 (2017)

10. Hanqing, G., Feng, K., Cao, Y.-L., Huang, M., Wu, Y.-H., Guo, Z.: Experimental and theoretical investigation of rotordynamic characteristics of a rigid rotor supported by an active bump-type foil bearing. J. Sound Vib. **466**, 115049 (2019)
11. Ha, D., Stolarski, T., Yoshimoto, S.: An aerodynamic bearing with adjustable geometry and self-lifting capacity. part 1: self-lift capacity by squeeze film. In: Proceedings of the Institution of Mechanical Engineers Part J-Journal of Engineering Tribology - PROC INST MECH ENG J-J ENG TR, vol. 219, pp. 33–39 (2005)
12. Belforte, G., Raparelli, T., Viktorov, V., Trivella, A.: Discharge coefficients of orifice-type restrictor for aerostatic bearings. Tribol. Int. **40**, 512–521 (2007)
13. Chang, S., Chan, C., Jeng, Y.: Numerical analysis of discharge coefficients in aerostatic bearings with orifice-type restrictors. Tribol. Int. **90**, 157–163 (2015)
14. Gao, S., Cheng, K., Chen, S., Ding, H., Fu, H.: CFD based investigation on influence of orifice chamber shapes for the design of aerostatic thrust bearings at ultrahigh speed spindles. Tribol. Int. **92**, 12 (2015)
15. Renn, J.-C., Hsiao, C.-H.: Experimental and CFD study on the mass flow-rate characteristic of gas through orifice-type restrictor in aerostatic bearings. Tribol. Int. **37**, 309–315 (2004)
16. Rowe, W.B.: Basic Flow Theory 1st(ed.), p. 25-48. Elsevier (2012)
17. Zhou, Y., Chen, X., Chen, H.: A hybrid approach to the numerical solution of air flow field in aerostatic thrust bearings. Tribol. Int. **102**, 04 (2016)
18. Larsen, J., Varela, A.C., Santos, I.: Numerical and experimental investigation of bump foil mechanical behaviour. Tribol. Int. **74**, 46–56 (2014)
19. Larsen, J., Santos, I., Hansen, A.: Experimental and theoretical analysis of a rigid rotor supported by air foil bearings. Mech. Ind. 09 (2014)
20. Larsen, J., Santos, I., von Osmanski, S.: Stability of rigid rotors supported by air foil bearings: comparison of two fundamental approaches. J. Sound Vib. **381**, 07 (2016)
21. Larsen, J., Nielsen, B., Santos, I.: On the numerical simulation of nonlinear transient behavior of compliant air foil bearings. In: Proceedings of the 11th International Conference on Schwingungen Rotierenden Maschinen, pp. 1–13 (2015)
22. Larsen, J., Santos, I.: On the nonlinear steady-state response of rigid rotors supported by air foil bearings - theory and experiments. J. Sound Vib. **346**, 03 (2015)
23. von Osmanski, S., Larsen, J., Santos, I.: A fully coupled air foil bearing model considering friction - theory & experiment. J. Sound Vib. **400**, 660–679 (2017)
24. Heinemann, S.T., Jensen, J.W., von Osmanski, S., Santos, I.F.: Numerical modelling of compliant foil structure in gas foil bearings: comparison of four top foil models with and without radial injection. J. Sound Vib. 117513 (2022)
25. von Osmanski, S., Santos, I.: Gas foil bearings with radial injection: multi-domain stability analysis and unbalance response. J. Sound Vib. **508**, 116177 (2021)
26. Heinemann, S.: Design of controllable segmented foil bearings based on multiphysics modelling techniques, Master's Thesis (2020)
27. Larsen, J., Santos, I.: Efficient solution of the non-linear reynolds equation for compressible fluid using the finite element method. J. Braz. Soc. Mech. Sci. Eng. **37**, 07 (2014)
28. von Osmanski, S., Larsen, J., Santos, I.: Modelling of compliant-type gas bearings: a numerical recipe. In: 13th International Conference on Dynamics of Rotating Machinery (SIRM 2019), pp. 13–27 (2019)
29. Hestmat, H., Walowit, J.A.: Analysis of gas lubricated compliant trust bearings. J. Lubr. Technol. **105**(4), 638–646 (1983)
30. Hestmat, H., Walowit, J.A.: Analysis of gas lubricated foil bearings. J. Lubr. Technol. **105**(4), 647–655 (1983)

31. Inman, D.J.: Engineering Vibration, 4th edn. Pearson, London (2014)
32. Pierart, F., Santos, I.: Active lubrication applied to radial gas journal bearings. part 2: modelling improvement and experimental validation. Tribol. Int. **96**, 12 (2015)
33. Annon. IEEE standard on piezoelectricity. ANSI/IEEE Std 176-19871 (1988)
34. Christensen, R., Santos, I.: Design of active controlled rotor-blade systems based on time-variant modal analysis. J. Sound Vib. **280**(3), 863–882 (2005)
35. Christensen, R., Santos, I.: Modal controllability and observability of bladed disks and their dependency on the angular velocity. J. Vib. Control **11**(6), 801–828 (2005)
36. Hendricks, E., Jannerup, O., Sørensen, P.H.: Linear Systems control, Deterministic and Stochastic Methods. Springer-Verlag, Berlin (2008)
37. Cook, R.D., Malkus, D.S., Plesha, M.E., Witt, R.J.: Concepts and Applications of Finite Element Analysis, 4th edn. Wiley, Hoboken (2001)
38. Dennis, J.J.E., Schnabel, R.B.: Numerical methods for unconstrained optimization and nonlinear equations. Society for Industrial and Applied Mathematics, SIAM (1983)
39. Brenan, K.E., Campbell, S.L., Petzold, L.R.: Numerical solution of initial-value problems in differential-algebraic equations. Society for Industrial and Applied Mathematics, SIAM (1996)
40. Morosi, S., Santos, I.: On the modelling of hybrid aerostatic-gas journal bearings. Proceedings of the Institution of Mechanical Engineers, Part J: Journal of Engineering Tribology, vol. 225, pp. 641–653 (2011)

Tilting-Pad Journal Bearing with Active Pads: A Way of Attenuating Rotor Lateral Vibrations

Heitor A. P. da Silva and Rodrigo Nicoletti[✉]

São Carlos School of Engineering, University of São Paulo,
São Carlos, SP 13566-590, Brazil
heitorantonio@usp.br, rnicolet@sc.usp.br

Abstract. Rotating machines can eventually operate under unpredicted or off-the-norm conditions, thus resulting in failure and significant economic losses. One way of reducing the consequences of such off-the-norm operation is the modification of the characteristics of the bearings that support the rotor. However, with conventional bearings, this is only possible through complete stops of the machine. The solution to this problem comes from the development of active bearings. In this sense, this work presents the proof-of-concept of a tilting-pad journal bearing whose pads are controlled by electromagnetic actuators. By mounting the electromagnetic actuators in the bearing casing, one can exert electromagnetic forces on the pads and make them change their angular position in relation to the bearing. Hence, one can modify the dynamic characteristics of the rotor-bearing system for desired and more appropriate values for a given operational condition. In this work, we present the system and we show the feasibility of reducing the lateral vibration response of a rotor mounted in a test bench via experimental results.

Keywords: active bearing · hydrodynamic lubrication · electromagnetic actuators · rotor dynamics · control systems

1 Introduction

The components of rotating machines of medium and large sizes can eventually fail under unpredicted or off-the-norm operating conditions, despite the tight design requirements defined in standards and norms. As a result, one has unprogrammed maintenance stops of the machine, thus leading to high economic losses for the industry. Examples of such losses can be easily found in the literature: dynamic instabilities in an off-shore gas compressor in the North Sea delayed production for six months [1]; the repair of seals in aircraft turbo engines due to excessive rotor vibrations represented 1% of the annual cost of the airline company [2]; an excessive lateral vibration of a 28 MW gas turbine led to such a catastrophic failure that machine repair was economically unfeasible [3]; excessive vibration and surge in a gas turbine of a thermoelectric power plant caused successive blackouts, thus harming the local population and resulting in penalty fees to the energy company [4], excessive vibration of a stage in a turbo-compressor caused the complete break of the stage that prevented repair [5].

In the energy prospection and generation sector, medium and large rotating machines, like turbo-generators, turbo-compressors, turbines, and pumps, are vital elements in the production process. Therefore, these machines shall present not only high performance but also high availability and safety [6]. A way of achieving such high performance and availability is the improvement of the tilting-pad journal bearings commonly used in these machines. For example, an off-shore natural gas compressor that presented high levels of sub-synchronous vibration in full load operating conditions had its tilting-pad journal bearings overhauled to solve the problem [7]. The problem of this kind of solution (alteration of bearing characteristics) is the need to completely stop the machine and production to change the machine component. Bearings are designed for delivering prescribed dynamic characteristics that, once defined, cannot be easily changed.

An alternative way of changing the dynamic characteristics of the bearings during the machine operation (without necessarily stopping the machine for a component exchange) is the introduction of sensing and actuating elements, which transforms the bearings into active bearings. The first idea of actively controlling vibration in rotating systems came with the advent of magnetic bearings [8]. However, the forces involved in large rotating machines require large and complex magnetic actuators to bear and control the motion of the rotor. Hence, the use of magnetic bearings only is not recommended for the application in large rotating machines. In this context, there comes the idea of a hybrid bearing, combining the actuation capacity of the magnetic bearing with the supporting capacity of the hydrodynamic bearing. The first ideas focuses on journal bearings supporting the rotor, with electromagnetic actuators embedded in the bearing casing [9–11]. By using the journal bearing as the supporting mechanism of the rotor (hydrodynamic lubrication), one can reduce the size of the electromagnetic actuators, because they only work as a control mechanism (not a supporting mechanism as it is in magnetic bearings). An additional advantage is the elimination of the backup bearing, which is a safety measure adopted in magnetic bearings in the case of electric failure.

The implementation of this solution to tilting-pad journal bearings came later, by embedding the electromagnetic actuators inside the tilting-pads [12] (Fig. 1). The experimental results of this kind of hybrid bearing showed an effective reduction of the rotor vibration levels, both in the time domain [13] and in the frequency domain [14], when the electromagnetic actuators are turned on. It was observed an 11% reduction of the system's resonance peak for the rotating system running at 600 rpm, and a reduction of 18% of the resonance peak for the system running at 1100 rpm, when adopting a Proportional-Derivative (PD) feedback control loop. The results also showed that electromagnetic actuators can be used to change the dynamic characteristics of the rotor-bearing system at any moment during the machine's operation.

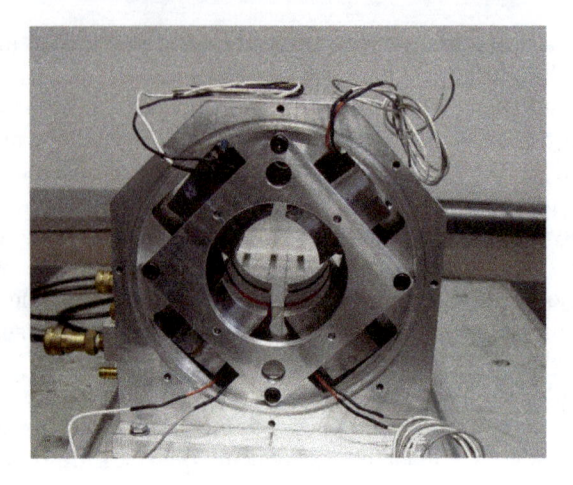

Fig. 1. Active tilting-pad journal bearing with embedded electromagnetic actuators in the pads.

However, this design solution also presented some drawbacks. The first drawback was the high moment of inertia of the pads due to the presence of the actuators, which reduced the natural frequencies associated with localized normal modes of the pads. For this reason, the dynamics of the pads (resonances) showed up in the frequency response of the rotor-bearing system. That complicated the control strategy because it resulted in spillover effects. The second drawback was the dynamic coupling that appeared between the electromagnetic forces (applied to the rotor) and the angular motion of the pads. When the actuator inside the pad was turned on, it attracted the rotor towards the pad but, by reaction, it also forced the pad towards the rotor, thus changing its angular position. This effect also complicated the performance of the control strategy.

To overcome these drawbacks, one modifies the position of the electromagnetic actuators inside the tilting-pad journal bearing, thus resulting in the tilting-pad bearing with active pads. The alternative is moving the actuators from the pads to the casing, facing the backside of the pads (Fig. 2). Hence, the electromagnetic force is no longer applied directly to the rotor but it is applied to the backside of the pads. The magnetic field of the actuators is trapped by the presence of the metallic pad and it does not affect the rotor directly. By doing this, one can alter the angular position of the pads, thus changing the hydrodynamic condition of the bearing, and consequently affecting the dynamic behavior of the rotor. This design solution solves the problem of the high inertia of the pads because they no longer have actuators inside them (their geometry is the same as that of common tilting-pad journal bearings). In addition, there is no dynamic coupling between the actuation forces on the rotor and the position of the pads, because the actuation forces act only on the pads (the action on the rotor is done by the hydrodynamic forces).

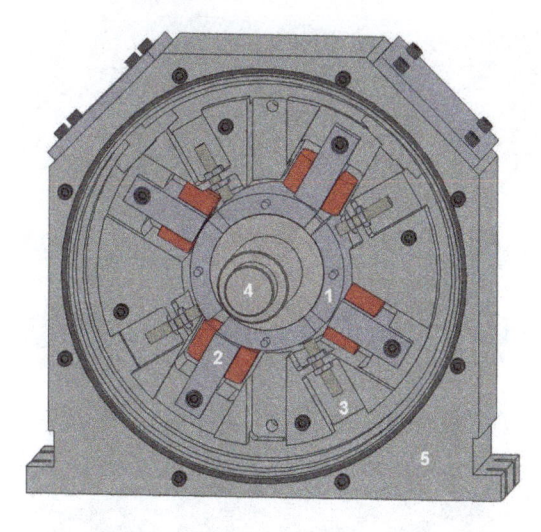

Fig. 2. Active tilting-pad journal bearing with active pads: 1) pads, 2) electromagnetic actuator, 3) proximity sensor, 4) rotor, 5) casing.

In this work, we present an experimental test rig designed to prove the concept of the tilting-pad bearing with active pads and we show experimentally the feasibility of reducing the lateral vibration response of the rotor. The test rig constrains the motion of the rotor to the vertical direction and experimental tests are performed in open-loop and closed-loop operating conditions.

2 The Proof-of-Concept Test Rig

The concept of the tilting-pad journal bearing with active pads is proved in a test rig (Fig. 3). In this test rig, the rotor displacements and the control action of the active bearing are limited to a single direction (vertical direction). Hence, the rotor (1) is mounted in a lever system with ball bearings (3) that restricts the motion of the rotor to the vertical direction but allows its rotation. The rotor is supported by the active bearing in the study (2) and it is connected to an electric motor (5) via a Cardan coupling (6), and the rotating speed of the rotor is controlled by a frequency inverter. Oil is supplied to the tilting-pad journal bearing by a hydraulic unit by pipelines connected to the bearing casing (7). The rotor and the pads are made of steel, whereas the bearing casing and the lever are made of aluminum. Angular contact ball bearings are used to hold the rotor in the lever and they also allow the reduction of mechanical clearances in the ball bearings. The pads are free to rotate around a pinned joint in the bearing casing. The parameter values of the test rig are listed in Table 1.

Proximity sensors (inductive analog transducers) are used to measure the motion of the pads and the vertical motion of the rotor. The adopted lubricant is ISO VG32, and the lubricant temperature is measured with Pt100 thermoresistance probes.

Fig. 3. Test rig of the tilting-pad journal bearing with active pads: 1) rotor, 2) active tilting-pad journal bearing, 3) rotor lever, 4) lever pivot, 5) electric motor, 6) Cardan coupling, 7) oil supply.

Table 1. Parameter values of the test rig.

parameter	value	unit	parameter	value	unit
rotor radius	40	mm	bearing assembled clearance	100	μm
pad radius	40.2	mm	bearing preload factor	0.5	—
pad length	60	mm	lever length (pivot-to-pivot)	270	mm
pad width	14	mm	rotor + lever mass	27.3	kg
pad aperture angle	80	degree	rotor + lever inertia	3.3	kg.m^2

The electromagnetic actuators of the active bearing have an E-shaped steel core where an AWG 21 copper wire is wound in 850 turns (Fig. 4). They are mounted in the backside of the pads, at a distance of 1.5 mm, and they apply an electromagnetic force that tends to tilt the pad when turned on. The actuators are connected to a drive that amplifies a low-power reference signal. By sending a reference signal with a voltage amplitude of V_{ref} to the drive, the drive splits the signal, sending to the actuator #1 (mounted on the upper pad) a high-

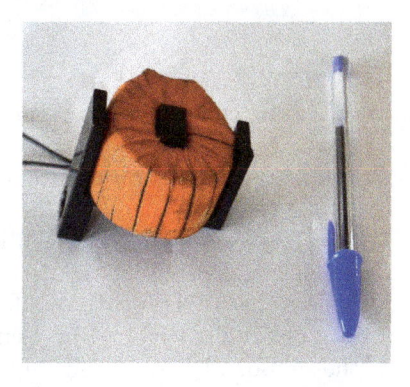

Fig. 4. Electromagnetic actuator of the tilting-pad journal bearing with active pads.

power voltage signal V_1 composed of the positive part of the reference signal, and sending to the actuator #2 (mounted on the bottom pad) a high-power voltage signal V_2 composed of the negative part of the reference signal in absolute value. In this case, the amplification factor of the drive is 2.4, and the maximum allowable electric current to the actuators is 5 A. Considering that the analog output port of the acquisition system can only supply reference signals within ±10 V, the maximum electric voltage applied to the actuators will be 24 V.

3 Experimental Results

The system is tested in two different operating conditions: in open-loop and closed-loop conditions. In the open-loop condition, there is no control loop feedback and the reference signal sent to the actuators' drive is independent of the motion of the rotor. In the closed-loop condition, control-loop feedback is implemented and the reference signal is proportional to the motion of the rotor by following a proportional-derivative (PD) control law. The bearing operating conditions in the tests are defined by the Sommerfeld number, which is a dimensionless parameter [15]:

$$S = \frac{\mu N L D}{W}\left(\frac{R}{C}\right)^2, \qquad (1)$$

where μ is the dynamic viscosity of the lubricant, N is the rotating speed of the rotor (in Hz), L is the bearing length, D is the bearing diameter ($D = 2R$), W is the load applied to the rotor, and C is the bearing machined clearance ($C = R_p - R$, where R_p is the pad radius).

3.1 Open-Loop Control

In the tests under open-loop operating conditions, no control loop feedback is implemented. A sinusoidal signal with amplitude A and angular frequency ω is sent to the actuators' drive as the reference signal:

$$V_{ref} = A\sin(\omega t). \tag{2}$$

The adopted amplitude of the reference signal was 5 V in all the presented results below.

Figure 5 presents the obtained experimental results for the rotating speeds of 594 rpm ($S = 0.35$) and 1190 rpm ($S = 0.66$). In these cases, the excitation frequency f 10 Hz 20 Hz ($\omega = 2\pi f$), respectively, which is similar to the rotor rotating frequency. The reference signal is sent to the actuators' drive at the time instant of 0.5 s. One can see that it is possible to affect the behavior of the rotor and impose to the rotor the desired motion by using the design solution proposed in this work. As the sinusoidal reference signal is sent to the drive and the drive sends the amplified signal to the actuators, the behavior of the pads is controlled by the actuators. Consequently, the hydrodynamic conditions in the bearing gap change, thus affecting the position of the rotor in the bearing (Fig. 5a). However, this effect decreases in the case of a higher rotating speed and higher excitation frequency (Fig. 5b). This is caused by two reasons: first, at higher rotating speeds, the equivalent stiffness of the lubricant film increases (a common hydrodynamic effect in lubricated bearings due to the higher fluid shear at high rotating speeds), thus requiring a higher force to move the rotor in a given amplitude; second, at higher excitation frequencies, the electromagnetic actuators tend to reduce their force, which is a known and expected behavior of inductive systems.

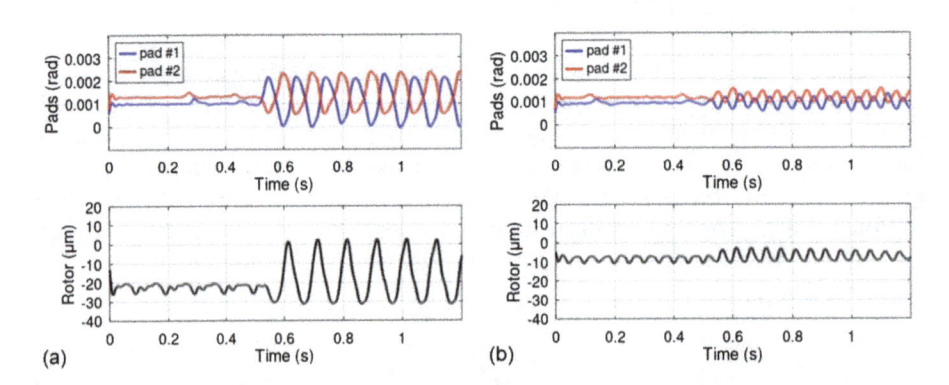

Fig. 5. Experimental results of the tilting-pad bearing with active pads under open-loop operating conditions (excitation frequency similar to the rotating frequency): (a) $S = 0.35$, 594 rpm, excitation frequency 10 Hz, (b) $S = 0.66$, 1190 rpm, excitation frequency 20 Hz.

Figure 6 presents the results for different excitation frequencies for the rotating speeds of 1190 rpm ($S = 0.66$) and 1790 rpm ($S = 0.97$). The results show that one can impose a high amplitude motion to the rotor with an excitation frequency 10 Hz, irrespective of the adopted rotating speed. By increasing the excitation frequency, the amplitude of the rotor motion decreases as expected.

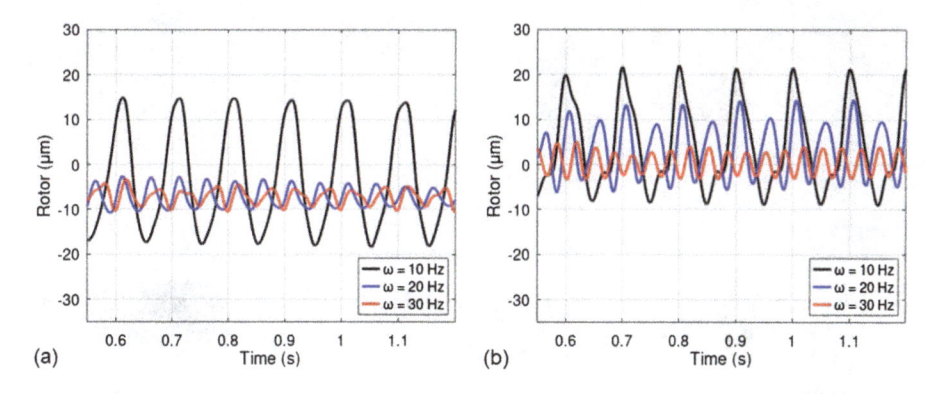

Fig. 6. Experimental results of the tilting-pad bearing with active pads under open-loop operating conditions (excitation frequency different from rotating frequency): (a) $S = 0.66$, 1190 rpm, (b) $S = 0.97$, 1790 rpm.

3.2 Proportional-Derivative Feedback Control

In the test under closed-loop operating conditions, one adopted the proportional-derivative (PD) control law:

$$V_{ref} = G_P(Y_R - Y_{ref}) + G_D(\dot{Y}_R - \dot{Y}_{ref}), \tag{3}$$

where G_P and G_D are the proportional and derivative gains, respectively, Y_R and $\dot{Y}_R$ are the vertical position and velocity of the rotor, and Y_{ref} and $\dot{Y}_{ref}$ are the reference values of the controller. In this case, Y_{ref} is the equilibrium position of the rotor in the bearing, and $\dot{Y}_{ref}$ is zero.

In the experimental tests, the rotor is subjected to an unbalance of 0.007 kg.m, and one adopted a proportional gain of 600 V/mm and a derivative gain of 7 V/(mm/s). Figure 7 presents the obtained experimental results when the controller is turned on at the time instant of 1 s. As one can see, the active bearing is effective in attenuating the vibration amplitude of the rotor, irrespective of the operating condition. At the operating condition of $S = 0.16$, there is a reduction of vibration amplitude of 26%; at the operating condition of $S = 0.35$, the reduction of vibration amplitude is 19%; and at the operating condition of $S = 0.64$, there is a reduction of vibration amplitude of 28%. In all these cases, the control signal amplitude value remained within ± 4 V.

This vibration attenuation of the rotor is an evidence of the higher damping introduced by the active bearing, which is a typical effect of controllers with derivative gains. However, the results in Fig. 7 also show that the active bearing moved the equilibrium position of the rotor when the controller is turned on at the time instant of 1 s. By looking at the results for the rotor position, one can observe that the controlled rotor (after $t = 1$ s) oscillates around an equilibrium position closer to the bearing center than that of the uncontrolled rotor (before $t = 1$ s). That is the result of the proportional gain, which is typically responsible for increasing the system's stiffness. Hence, by turning on the controller,

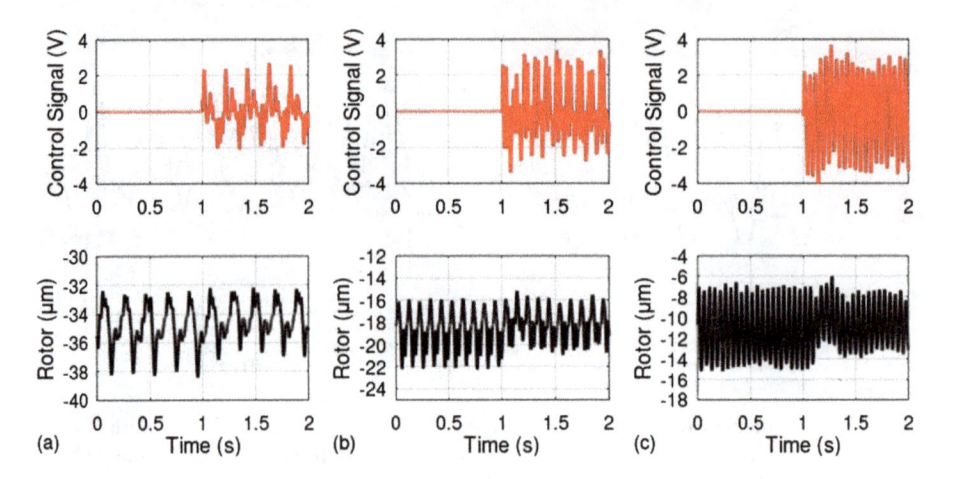

Fig. 7. Experimental results of the tilting-pad bearing with active pads under closed-loop operating conditions (control after $t = 1$ s): (a) $S = 0.16$, 296 rpm, (b) $S = 0.35$, 593 rpm, (c) $S = 0.64$, 1190 rpm.

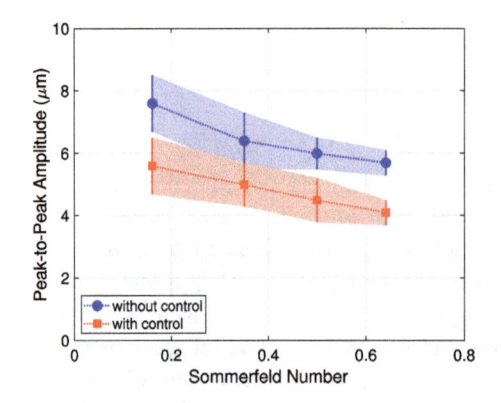

Fig. 8. Peak-to-peak vibration amplitude of the rotor as a function of the Sommerfeld number with and without control (uncertainty shown as shaded area).

the rotor-bearing stiffness increases, thus shifting the rotor equilibrium position towards the bearing center.

A summary of the obtained results is shown in Fig. 8, where peak-to-peak rotor vibration amplitude is plotted as a function of the Sommerfeld number for four different operating conditions. The shaded area in the figure gives an idea of the uncertainty of the amplitude measured during operation. One can see that vibration amplitude is very low (below 8 μm peak-to-peak), due to the low unbalance level adopted. However, the active bearing with a PD-controller managed to reduce the vibration amplitude, as evidenced in Fig. 8.

4 Conclusion

This work presents experimental results of the proof-of-concept of a tilting-pad journal bearing with active pads. The experimental results obtained in a laboratory test rig shows the feasibility of controlling the movements of the rotor by using the proposed design solution for the active bearing. In the open-loop operating conditions, it is verified that higher actuation forces are achieved at lower vibration frequencies due to the inductive nature of the electromagnetic actuators. However, in the closed-loop operating conditions, it is confirmed that the actuation system is capable of attenuating the rotor vibrations irrespective of the considered rotating speeds. Indeed, the highest vibration amplitude reduction (28%) is obtained at the highest rotating speed tested (1190 rpm). These results show that the proposed active bearing represents a promising design solution to the control and attenuation of rotor lateral vibrations in rotating machines, enabling the alteration of rotor dynamic characteristics during the machine operation.

Funding. This project was supported by the Brazilian research foundations Conselho Nacional de Desenvolvimento Científico e Tecnológico (Grant no.: 304212/2021-0), Fundação de Amparo à Pesquisa do Estado de São Paulo (Grant no.: 2019/23220-2), and Coordenação para o Aperfeiçoamento de Pessoal de Nível Superior (Grant no.: 88887.465094/2019-0).

References

1. Ehrich, F., Childs, D.: Self-excited vibration in high-performance turbomachinery. Mech. Eng. **106**(5), 66–79 (1984)
2. Bartha, A.R.: Dry friction backward whirl of rotors. Swiss Federal Institute of Technology, Zurich, PhD Thesis (2000)
3. Farrahi, G.H., Tirehdast, M., Abad, E.M.K., Parsa, S., Motakefpoor, M.: Failure analysis of a gas turbine compressor. Eng. Fail. Anal. **18**(1), 474–484 (2011)
4. Fontes, C.H., Budman, H.: A hybrid clustering approach for multivariate time series: a case study applied to failure analysis in gas turbine. ISA Trans. **71**(2), 513–529 (2017)
5. Azevedo, T.F., Cardoso, R.C., Silva, P.R.T., Silva, A.S., Griza, S.: Analysis of turbo impeller rotor failure. Eng. Fail. Anal. **63**, 12–20 (2016)
6. Pereira, L.P., Machado, M.M., Valland, A., Manguinho, D.A.P.: Failure intensity of off-shore power plants under varying maintenance policies. Eng. Fail. Anal. **97**, 434–453 (2019)
7. Li, J., Choudhury, P., Taques, R.: Seal and bearing upgrade for eliminating rotor instability vibration in a high pressure natural gas compressor. In: ASME Turbo Expo 2002: Power for Land, Sea, and Air, pp. GT2002-30635. ASME, Amsterdam (2002)
8. Schweitzer, G., Maslen, E.H.: Magnetic Bearings. Springer Verlag, Heidelberg (2009). https://doi.org/10.1007/978-3-642-51724-2
9. Mecanique Magnetique SA: Hybrid fluid bearing with stiffness modified by electromagetic effect. Patent 4827169, 2 May 1989

10. Mechanical Tech Inc.: Active magnetic bearing device for controlling rotor vibrations. Patent US5059845, 22 October 1991
11. Eastman Kodak Co.: Hydrodynamic fluid bearing with electromagnetic levitation. Patent GB2296945, 17 July 1996
12. Nicoletti, R.: Mancal hidrodinâmico segmentado ativo com sapatas móveis magnéticas (Active tilting-pad journal bearing with magnetic pads). Patent PI 0700697-7, 30 September 2008
13. Moraes, D.C., Nicoletti, R.: Hydrodynamic bearing with electromagetic actuators: rotor vibration control and limitations. In: International Conference on Noise and Vibration Engineering, Proceedings of ISMA, pp. 1–7. KU Leuven, Leuven (2010)
14. Viveros, H.P., Nicoletti, R.: Lateral vibration attenuation of shafts supported by tilting-pad journal bearing with embedded electromagnetic actuators. J. Eng. Gas Turbines Power **136**(4), 042503 (2014)
15. Someya, T.: Journal-Bearing Databook. Springer Verlag, Berlin, Heidelberg (1989). https://doi.org/10.1007/978-3-642-52509-4

Studying the Effect of Viscous Friction Minimization in Actively Lubricated Journal Hybrid Bearings

Denis Shutin[✉], Leonid Savin, and Yuri Kazakov

Orel State University n.a. I.S. Turgenev, Komsomolskaya str. 95, 302026 Orel, Russia
`rover.ru@gmail.com`

Abstract. This work considers the effect of minimization of viscous friction in actively lubricated journal hybrid bearings by choosing the optimal setpoint of the rotor position. The configuration of the pressure distribution and the gap shape in the bearing can be adjusted so that the resulting viscous forces ensure a minimum of friction under full film lubrication conditions. In general case, the rotor position corresponding to the minimum viscous friction is different from its natural position under similar conditions in a passive bearing, when the lubricant is supplied through a common manifold under constant pressure. The results of the experimental studies including the direct measurement of friction torque in the bearings agree well with the results of numerical simulation. The severity of friction reduction effect and variation of location of the minimum friction point under different parameters of the rotor-bearing system are theoretically studied. Stability of the rotor initially set by the control system to the minimum friction position under complex loading conditions is considered. Two types of controllers, a conventional P-controller and a more advanced adaptive P-controller, are tested for this purpose. The obtained results allow to draw a conclusion about how much this effect should be considered when designing machines with active hybrid bearings.

Keywords: Active Lubrication · Active Fluid Film Bearings · Friction Reduction · Rotary Machines

1 Introduction

Viscous friction is a significant factor leading to power losses in fluid film bearings. The current trend towards energy efficiency of machines influences their design. This has led to an increase in research in the field of optimizing the parameters of bearings, including reducing friction. Influence of design parameters on energy efficiency was studied, for example, in [1, 2]. Increasing efficiency of optimization procedures can be achieved by using the corresponding computing tools. The paper [3] considers the use of genetic algorithms for solving the mentioned problem. Friction issues are often considered together with wear processes [4]. There are few works that are devoted to the study of active friction reduction. Murashima et al. in [5] achieved a reduction in viscous friction losses by adjusting the shape of the bearing surface. Khatri et al. [6] analyzed viscous friction in an adjustable bearing with electrorheological fluid lubrication. However, if it comes to classical design, then in this case there are restrictions on achieving energy efficiency. The use of active fluid friction bearings makes it possible to overcome this threshold.

Number of studies in the field of active fluid friction bearings began in the 2000s, in recent years this topic has gained great popularity. Three main directions for the development of such bearings can be distinguished: minimizing friction, minimizing rotor vibrations magnitudes, and minimizing the vibrations of the whole rotary machine [7, 8]. Authors pay much attention to the design of active bearings, and the control systems are often considered superficially. Variable geometry of a bearing is often considered as an approach to building adjustable rotor-bearing systems [7, 9, 10]. Another possible approach to control the bearing parameters is active lubrication. Such bearings have been studied by Santos et al. [11]; Breńkacz et al. presented a number of papers on this topic in their review [7].

Friction reduction in actively lubricated conical bearings was considered by Kazakov et al. [12]. Friction torque is adjusted by varying the bearing clearance by controlled lubricant supply pressure. The available techniques of adjusting friction in actively lubricated journal bearings differ from ones for conical bearings. The correlation between the rotor position in an actively lubricated journal bearing and the viscous friction has been theoretically found and described in [13].

The present paper develops the results presented in [13], including its experimental verification and giving more detailed study of the friction reduction effect at different parameters of a rotor-bearing system.

2 Experimental Study

The experimental study has been conducted to confirm the correctness of the theoretical results obtained before. The considered actively lubricated journal bearing is shown in Fig. 1. The bearing includes a brass bushing with four lubricant supply orifices. The bearing diameter is of 40 mm, the length is of 62 mm, the radial clearance is of 130 μm. The feeding orifices have length of 11 mm and diameter of 2 mm. They supply the bearing with pressurized water. The lubricant pressure is adjusted by the servo valves Sv_i separately in each lubricant supplying channel.

The bearing was installed in a test rig shown in Fig. 2. The presented rotor-bearing system includes two hybrid bearings, one operating in passive mode (motor side), and another in active mode (free side). The electric motor is connected to the shaft through a torque sensor and a gear coupling. Two displacement sensors are arranged perpendicularly at the active bearing end of the shaft.

The rotor full mass is of 3 kg, a half of its weight is put on the active bearing. The torque sensor measures the friction torque generated by the viscous forces in two bearings. Although the sensor measures the sum of the frictional torques in the two bearings, the friction in the passive bearing does not change much even when the lubrication regime and shaft position in the active bearing change. Thus, the measured variations in the friction torque are mainly provided by the operation of the active bearing considered in this study.

The control system obtains the actual rotor position in the active bearing from two displacement sensors. The hydrodynamic part of the bearing force is not adjustable,

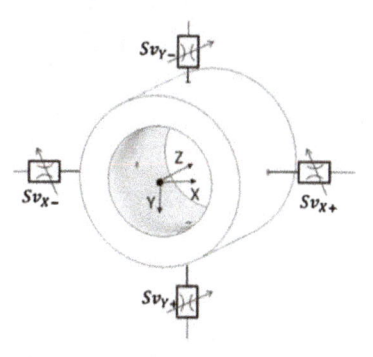

Fig. 1. Actively lubricated journal hybrid bearing

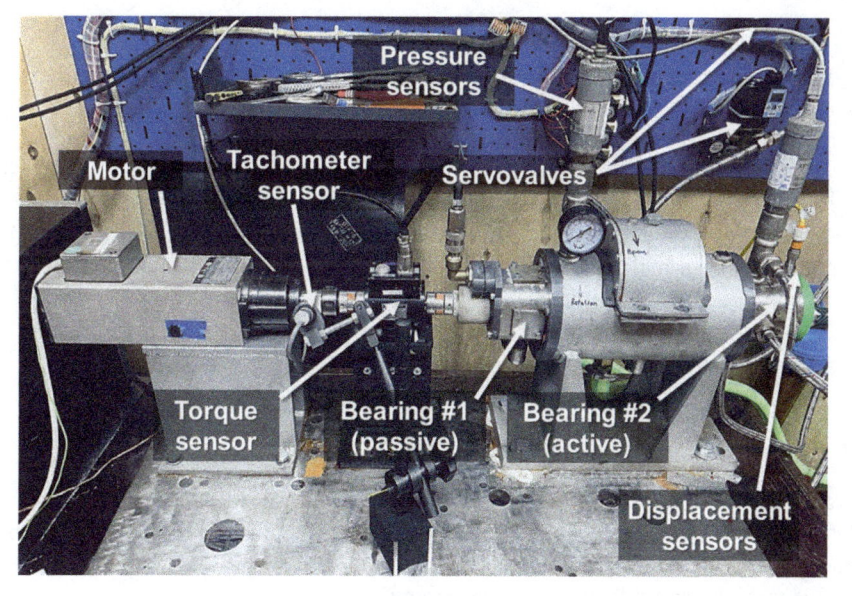

Fig. 2. Test rig

while the hydrostatic part is influenced by the servo valves. The structure and operational principle of the control system is described in more detail in [13].

The design of the experiment corresponds to the design of the theoretical study on identification of viscous friction as a function of a setpoint in the control system in [13]. However only 25 setpoint values have been tested in order to reduce the experiment time. The tested setpoints are distributed within the area of 0.6 of the bearing clearance h_0 around the bearing center with the angle step of 45°. Rotation speed of the shaft was of 1500 rpm. The signal of the torque sensor was averaged for the period of at least 60 s for each point. As a result, the range of values of the friction torque in the bearing was obtained. The experimentally obtained functional is presented in Fig. 3 together with the theoretically calculated one for a rotor-bearing systems with the corresponding parameters.

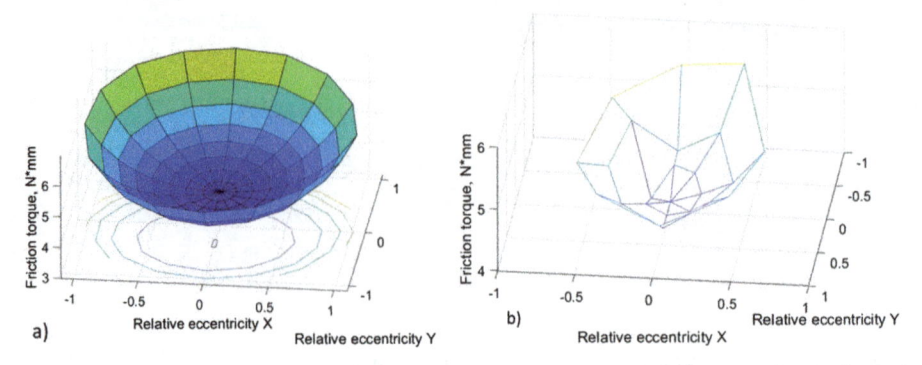

Fig. 3. The dependence of the viscous friction torque on the active bearing setpoint obtained: a) theoretically and b) experimentally

Despite the presence of an error in the torque sensor measurements due to the relatively low torque value, the experimental result qualitatively corresponds to the theoretical one based on the numerical solution of the modified Reynold's equation [13]. In both cases, there is a minimum of the functional near the bearing center of the bearing, slightly shifted in the positive direction of the X axis. The range of variation of the friction torque in the experiment also corresponds to the theoretical result, taking into account friction in the passive bearing. This confirms the adequacy of the theoretical model and the reliability of the results shown below.

3 Theoretical Study

3.1 Rotor Position with Minimum Friction

Viscous friction in a bearing is related to the pressure distribution in the lubricant film. The pressure distribution is influenced by many parameters of the rotor system, both structural and operational. Their change can also lead to a change in the functional of viscous friction from the rotor position setpoint.

Adjustable rotation speed is quite common in different types of machines. Figure 4 shows how the position of the minimum friction point changes in the shaft speed range from 500 to 10000 rpm. Note, that in this and further calculations, the parameters of the simulated rotor-support system correspond to those described in Sect. 2, excluding these which values are directly indicated in the description.

As can be seen from the results, the minimum friction point of is approximately at the level of the center of the bearing in vertical direction in the entire considered speed range. The point also shifts horizontally with an increase in the shaft speed from an eccentricity of about 0.35 to an almost central position.

Several more notes should me made regarding the result obtained. Firstly, the minimum friction point is determined with a certain error since the calculations were made only at the grid nodes. Secondly, the friction value increases insignificantly near the minimum point. As illustrated in Fig. 5, it varies by no more than 2% at a distance of $0.2 \cdot h_0$ at 1000 rpm. Accordingly, with relatively small deviations from the minimum point, the

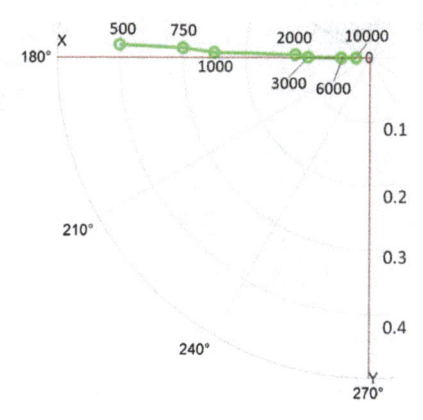

Fig. 4. Dependence of the rotor position with the minimum of friction on the rotation speed

effect of friction reduction practically does not decrease. This reduces the requirements for precise positioning of the rotor in the active bearing when the friction minimization should be obtained. The severity of friction reduction in the bearing is considered in more detail in the next section.

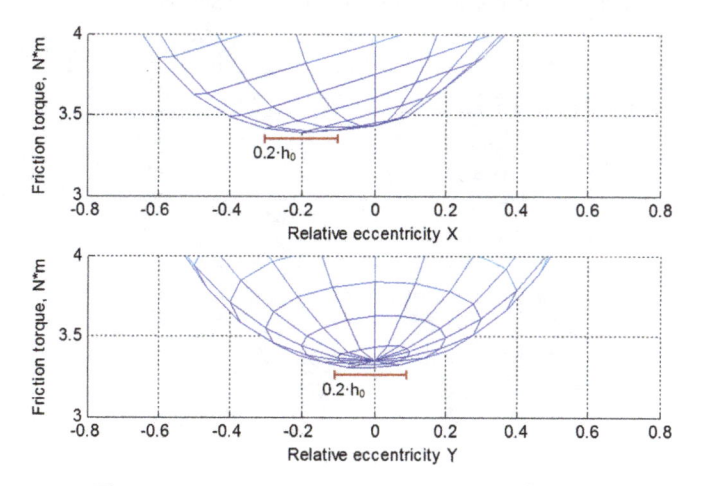

Fig. 5. Friction torque change near the minimum point

3.2 Severity of the Friction Reduction Effect

The quantitative assessment of the considered effect is most evident when comparing the friction in a passive bearing and in an active one, when the rotor is set to the calculated minimum friction point. The effect of several factors such as rotation speed and bearing clearance on viscous friction reduction is shown in Fig. 6.

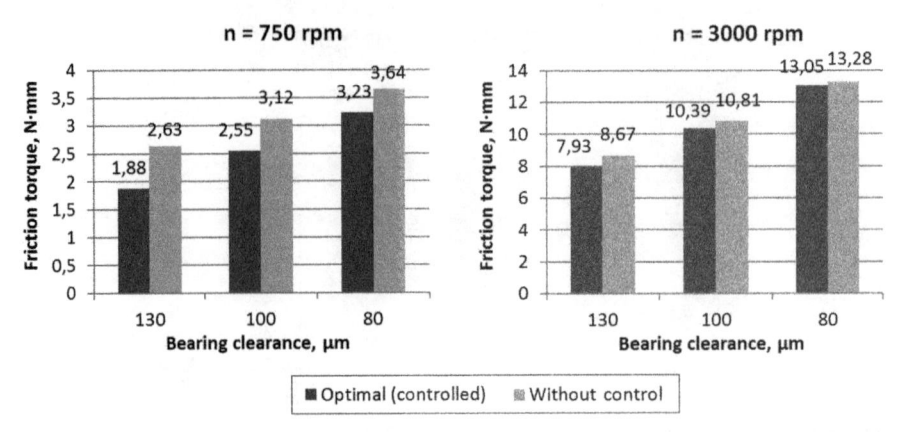

Fig. 6. Severity of friction reduction at different rotation speeds and bearing clearances

The greatest reduction in friction is observed for bearings with larger clearance. The severity of the effect decreases with an increase in rotation speed. For a bearing with a clearance of 130 μm, the predicted reduction in friction can be up to 28% at 750 rpm.

Since the function of the friction torque from the setpoint tends to increase with increasing rotor position eccentricity, the reduction in friction for systems with different rotor masses, i.e., different Sommerfeld numbers [14], was additionally analyzed. Figure 7 shows that greater friction reduction, up to 3 times or more, can be achieved for heavier rotors with a more eccentric natural position.

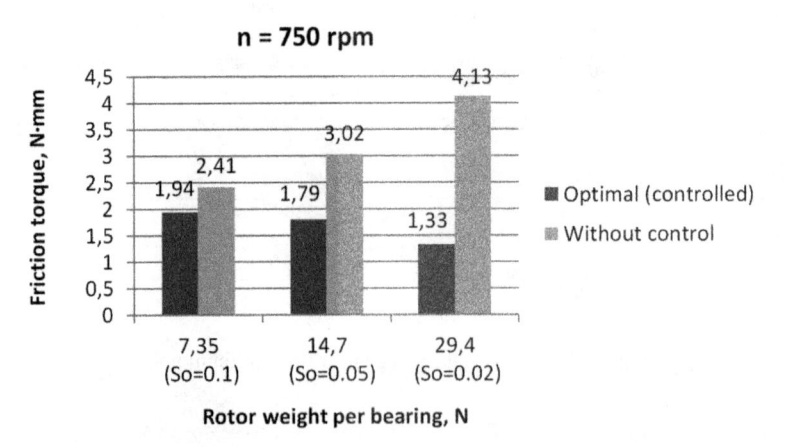

Fig. 7. Severity of friction reduction at different loads

3.3 Stability of Rotor Motion

As the results above show, the minimum value of viscous friction in an actively lubricated hybrid bearing can be achieved by holding the shaft close to the bearing center. In a plain bearing, a more centered saft position leads to decrease in hydrodynamic forces and overall load capacity, and can also adversely affect the stability of the rotor motion.

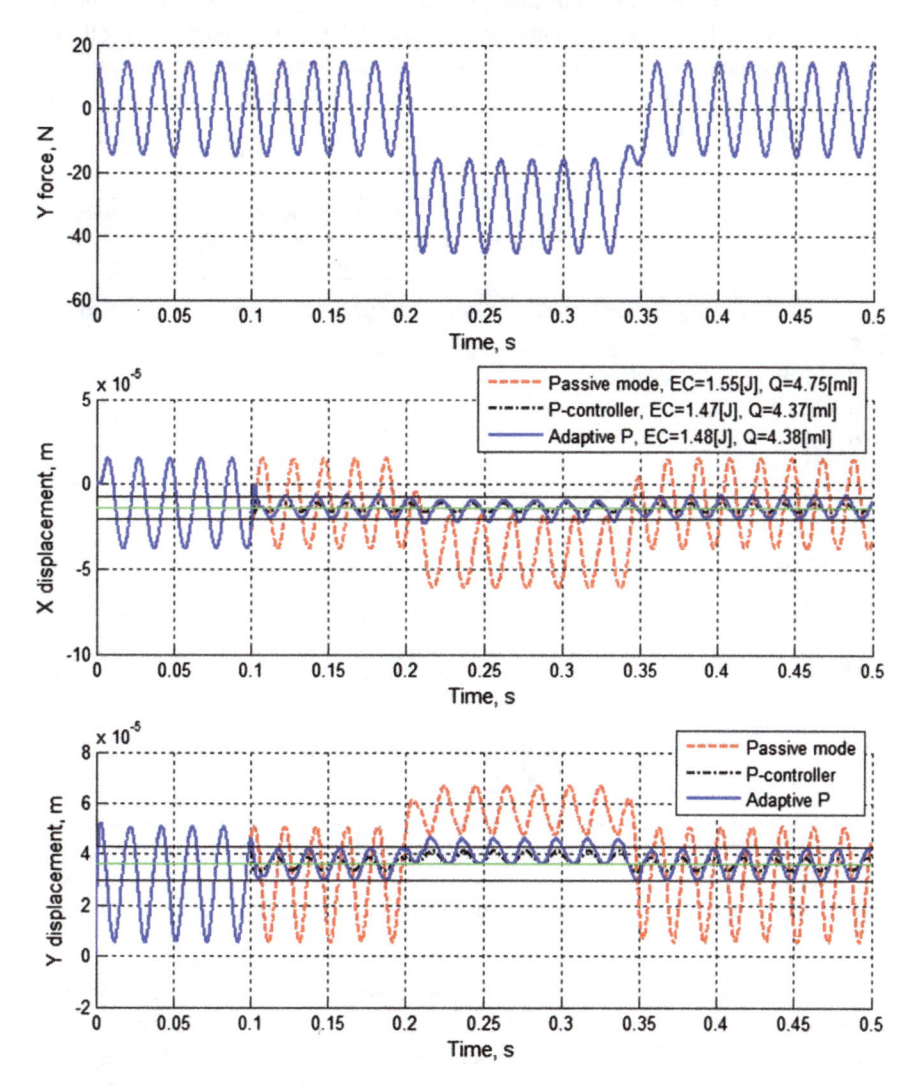

Fig. 8. Operation of the rotor-bearing system with the setpoint near the natural shaft position

Operation of a rigid rotor on actively lubricated hybrid bearings under combined load was simulated to test the stability of the system. The two scenarios were implemented with the simulation model, one with the setpoint near the natural shaft position (Fig. 8), and another with the setpoint at the minimum friction point near the bearing center (Fig. 9). In both cases the load applied to the rotor included the imbalance forces of value by the order of the rotor weight (15 N per bearing), and an additional force of value of two rotor weights applied vertically for 150 ms in the middle of the simulated period. For each case the motion of rotor in passive bearings is compared to its motion in active bearings with the two types of controllers, a conventional P-controller and an adaptive P-controller with adjustable proportional gain [15]. Operation of the rotor system with active bearings begins in passive mode at 3000 rpm, and the controllers are turned on after the first 100 ms. Also, for each case an overall energy consumption due to work of viscous friction forces (the 'EC' parameter) and an overall lubricant flow (the 'Q' parameter) are calculated.

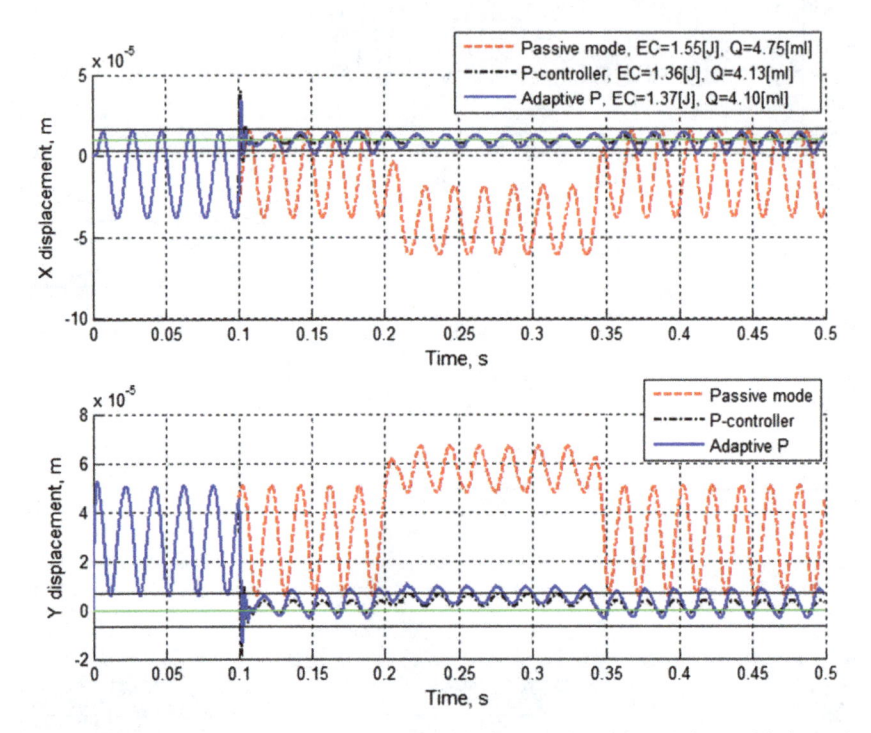

Fig. 9. Operation of the rotor-bearing system with the setpoint at the minimum friction point

The simulation results show that in the both cases the rotor motion remains stable. Active bearings reduce several times the magnitude of the rotor oscillations caused by imbalance forces and the shaft displacement due to the applied radial force. As for friction, the energy spent on it is reduced by 5% compared to the passive system, even if the setpoint is chosen near the natural shaft position. If the setpoint is chosen at the minimum friction point, the energy consumption is reduced by more than 12% compared to a passive bearing. Lubricant consumption is thus reduced by 8 and 13%, respectively.

The difference between the effect of conventional and adaptive P-controllers is negligible in almost all respects. However, the adaptive P-controller constantly holds the oscillation amplitude close to the specified limits. This helps avoiding overdamping, and also unnecessary work and wear of servo valves, thereby saving their resource.

4 Summary and Conclusions

The present work describes in more detail the previously identified effect of reducing viscous friction in actively lubricated hybrid bearings. The friction is reduced due to active rotor positioning and the corresponding changes in the pressure distribution in the bearing. The influence of this effect on energy and dynamic parameters of rotor-bearing systems was studied theoretically using the verified simulation model. The following conclusions can be made.

1. The conducted experimental studies confirm the presence of a minimum of viscous friction in an active journal hybrid bearing, which was previously discovered theoretically. The minimum can be achieved by holding the shaft position near the calculated point by the bearing's control system.
2. The value of the required setpoint, as well as achievable degree of reduction of viscous friction, depends on the operational and structural parameters of the rotor-bearing system, such as bearing clearance, specific load, rotation speed. As the rotation speed increases, the required setpoint moves closer to the center of the bearing.
3. The effect of viscous friction reduction is more pronounced at lower rotation speeds and larger bearing clearances. It is also much more pronounced for heavier rotors with a smaller value of the Sommerfeld number. In such systems, viscous friction can be reduced by more than 3 times compared to the performance of a similar passive bearing.
4. Despite the weakening of the hydrodynamic forces with a more centered shaft position in a plain journal bearing, rotor system with actively lubricated hybrid bearings remains stable due to the controlled hydrostatic action. In this case a reduction in viscous friction, vibrations magnitude and lubricant consumption can be achieved.

Acknowledgements. The study was supported by the Russian Science Foundation grant No. 22-19-00789, https://rscf.ru/en/project/22-19-00789/.

References

1. El-Sherbiny, M., Salem, F., El-Hefnawy, N.: Optimum design of hydrostatic journal bearings part II. Minimum power. Tribol. Int. **17**, 162–166 (1984). https://doi.org/10.1016/0301-679X(84)90009-4
2. Doshi, N., Bambhania, M.: Optimization of Film Thickness for Hydrostatic Circular Pad Bearing Used in V-25 vertical Turning Machine, pp. 2321–5747 (2013)
3. Saruhan H.: Minimization of power loss in hydrodynamic bearings design using the genetic algorithm. Süleyman Demirel Üniversitesi Fen Bilim. Enstitüsü Derg. 9–3 (2009). https://doi.org/10.19113/sdufbed.09574
4. Hu, B., Zhou, C., Wang, H., Chen, S.: Nonlinear tribo-dynamic model and experimental verification of a spur gear drive under loss-of-lubrication condition. Mech. Syst. Signal Process. **153**, 107509 (2021). https://doi.org/10.1016/J.YMSSP.2020.107509
5. Murashima, M., et al.: Active friction control in lubrication condition using novel metal morphing surface. Tribol. Int. **156**, 106827 (2021). https://doi.org/10.1016/j.triboint.2020.106827

6. Khatri, C.B., Sharma, S.C.: Analysis of textured multi-lobe non-recessed hybrid journal bearings with various restrictors. Int. J. Mech. Sci. **145**, 258–286 (2018). https://doi.org/10.1016/J.IJMECSCI.2018.07.014
7. Breńkacz, Ł., Witanowski, Ł., Drosińska-Komor, M., Szewczuk-Krypa, N.: Research and applications of active bearings: a state-of-the-art review. Mech. Syst. Signal Process. **151** (2021). https://doi.org/10.1016/j.ymssp.2020.107423
8. Qin, F., Li, Y., Qi, H., Ju, L.: Advances in compact manufacturing for shape and performance controllability of large-scale components-a review. Chin. J. Mech. Eng. (Engl. Ed.) **30**, 7–21 (2017). https://doi.org/10.3901/CJME.2016.1102.128
9. Chasalevris, A., Dohnal, F.: Improving stability and operation of turbine rotors using adjustable journal bearings. Tribol. Int. **104**, 369–382 (2016). https://doi.org/10.1016/J.TRIBOINT.2016.06.022
10. Pai, R., Parkins, D.W.: Performance characteristics of an innovative journal bearing with adjustable bearing elements. J. Tribol. **140** (2018). https://doi.org/10.1115/1.4039134
11. Haugaard, A.M., Santos, I.F.: Multi-orifice active tilting-pad journal bearings-Harnessing of synergetic coupling effects. Tribol. Int. **43**, 1374–1391 (2010). https://doi.org/10.1016/j.triboint.2010.01.009
12. Kazakov, Y.N., Kornaev, A.V., Shutin, D.V., Li, S., Savin, L.A.: Active fluid-film bearing with deep q-network agent-based control system. J. Tribol. **144** (2022). https://doi.org/10.1115/1.4053776
13. Shutin, D., Savin, L., Polyakov, R.: Influence of a control system in an active journal hybrid bearing on the energy parameters of its operation. Int. J. Energy Environ. **11**, 38–41 (2017)
14. Edwards, K.L.: Mechanical Engineering Design, 5th edn. In: Shigley, J.E., Mischke, C.R. (eds.) McGraw-Hill, Boston (1989). https://doi.org/10.1016/0261-3069(94)90047-7. ISBN 0-07-100607-9, 779 pp, £17.95. Mater. Des. 15, 116–117 (1994)
15. Shutin, D., Polyakov, R.: Adaptive nonlinear controller of rotor position in active hybrid bearings. In: 2016 2nd International Conference on Industrial Engineering, Applications and Manufacturing (ICIEAM), - Proceedings (2016). https://doi.org/10.1109/ICIEAM.2016.7910935

On the Feedback Control of a Rotor System with Active Flexible Bearings

Alexander Bitner[✉] and Carsten Proppe

Karlsruhe Institute of Technology, 76131 Karlsruhe, Germany
`alexander.bitner@kit.edu`
`https://www.itm.kit.edu/english/dynamik/index.php`

Abstract. Passive and active vibration control of rotor-bearing systems is frequently employed in order to suppress rotor vibrations due to instability, unbalance or external excitation. The use of non-circular bearings or tilting-pad bearings, for example, is a classical approach in the field of passive vibration control. In current developments, attempts are being made to optimize passive control or to add active control mechanisms to the rotor-bearing system. The latter includes the active flexible bearing (AFB) under investigation.

A simple model of a rotor-bearing system is introduced, consisting of a balanced rigid rotor supported in fluid-film bearings with elastically deformable outer bearing rings. The outer bearing ring is modeled as a thin and inextensible curved Euler-Bernoulli beam. Following an active vibration control (AVC) scheme, linear actuators are used to apply control forces to the elastic bearing ring, allowing its original circular shape to be changed to a non-circular one.

For non-stationary bearing rings, the effect of an alternating clearance geometry on the onset of whirl motion of the rotor is investigated during run-up simulations.

Keywords: vibration control · instability · flexible fluid-film bearing

1 Introduction

Passive and active vibration control of rotordynamic systems [1] aim to attenuate or suppress rotor vibrations due to unbalance, external excitation or loss of stability. The use of two-lobe bearings [2], pressure-dam bearings [3,4] or tilting-pad bearings [5], in order to suppress oil-whirl instability and increase the operation range of the rotor system, is a classical approach in the field of passive vibration control (PVC).

The use of active vibration control (AVC) in rotordynamic systems [6–13] has attracted increased research interest in recent decades. In this context, a rotordynamic system is extended by an active vibration control. An AVC system, which contains both sensor and actuator components, is able to monitor the state of its rotordynamic system and, if necessary, to regulate the rotordynamic behavior following a given control strategy. Typically, AVC systems can be attached to rotors [6], dampers [7], bearings [8–13] and seals [14], or other machine elements.

Krodkiewski and Sun [8,9] investigated multi-bearing rotor systems with an active journal bearing and its dynamic properties. The partially flexible bearing sleeve can be considered as a new feature of the proposed active journal bearing. The deformation of the flexible sleeve can be changed hydraulically by adjusting the chamber pressure. Therefore the clearance geometry and thus the dynamic properties of the rotor system can be controlled during operation of the rotor.

Carmignani et al. [10] modified a Bently Nevada Rotor Kit with an auxiliary fluid-film bearing which was actuated by piezoelectric actuators. The research objective was to verify the feasibility of active vibration control and the possibility of vibration reduction of a flexible, unbalanced rotor. Experimental tests successfully showed the reduction through the active vibration control by actuating the auxiliary bearing.

Tuma et al. [12] designed a test rig of a rotor-bearing system extended by an AVC system actuating the positions of the movable bearings. The test rig can be used to demonstrate that by actuating the bearing positions, it is possible to achieve a suppression of oil whirl instability to a certain extent. The bearings are suspended and moved by piezoelectric actuators according to a proportional feedback control reducing the difference between predefined and actual journal positions. Thus, not only the threshold of instability can be increased but a predefined journal position can be realized within the accuracy of microns.

The vertical rotor system under investigation is shown in Fig. 1. The position of the rotor's center of mass S_R is described by the coordinates x_R and z_R with respect to the inertial reference frame $\left(e_x^I, e_y^I, e_z^I\right)$. The rotation and angular speed of the rotor is given by the angle φ_R and its time derivative $\dot{\varphi}_R$, respectively. The vertical rigid rotor is supported by active flexible bearings (AFB) as shown in Fig. 1. An AFB consists of an elastically deformable outer bearing ring supported by two actuators and equipped with one displacement sensor. Depending on the rotor's lateral displacement z_R the control forces F_1 and F_2 exerted by the actuators will deform the bearing ring and influence the clearance geometry (see Fig. 1).

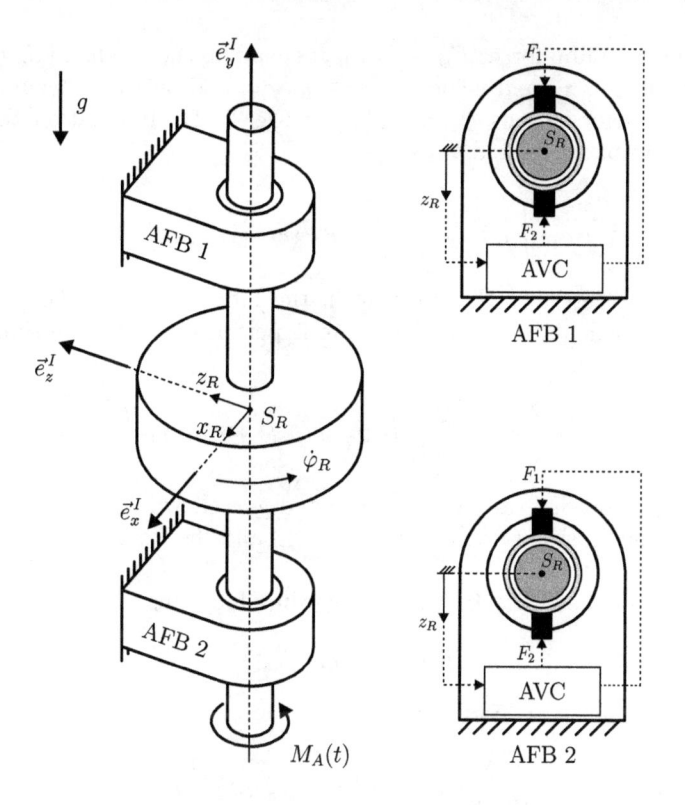

Fig. 1. Vertical Rotor system with fluid-film bearings

2 Modeling of the Vertical Rotor-Bearing System

2.1 Vertical Rigid Rotor and Fluid-Film Bearings

The dynamics of the vertical rotor system is governed by the equations of motion

$$J_R \ddot{\varphi}_R(t) = M_A(t) \tag{1}$$

$$M_R \ddot{x}_R(t) = 2F_{Bx}\left(x_R, z_R, \dot{x}_R, \dot{z}_R, \dot{\varphi}_R, w, \frac{\partial w}{\partial \theta}\right) \tag{2}$$

$$M_R \ddot{z}_R(t) = 2F_{Bz}\left(x_R, z_R, \dot{x}_R, \dot{z}_R, \dot{\varphi}_R, w, \frac{\partial w}{\partial \theta}\right) \tag{3}$$

in which the degrees of freedom are given by the coordinates x_R, z_R, φ_R and the system parameters by the rotor mass M_R as well as the mass moment of inertia J_R. The rotor system is driven by the torque

$$M_A(t) = \begin{cases} M_{A0}, & 0 \le t \le \Delta t \\ 0, & \text{otherwise} \end{cases}. \tag{4}$$

The nonlinear oil film forces F_{Bx} and F_{Bz} depend on the one hand on the rotor's displacements x_R, z_R and velocities $\dot{x}_R$, $\dot{z}_R$, $\dot{\varphi}_R$, and on the other hand on the bearing ring deformations w and $\frac{\partial w}{\partial \theta}$ (see Fig. 2). By integration the oil film forces [15] can be calculated

$$\begin{pmatrix} F_{Bx}(t) \\ F_{Bz}(t) \end{pmatrix} = -\int_0^B \int_0^{2\pi} p(\theta, y, t) \begin{pmatrix} \sin\theta \\ \cos\theta \end{pmatrix} r d\theta dy \tag{5}$$

where $p(\theta, y, t)$ describes the pressure in the lubrication gap. For the sake of simplicity, the calculation of the pressure is performed using the short bearing approximation

$$\frac{\partial^2}{\partial y^2} p(\theta, y, t) = \frac{\mu}{h^3(\theta, t)} \left(12 \frac{\partial}{\partial t} h(\theta, t) + 6\dot{\varphi}_R(t) \frac{\partial}{\partial \theta} h(\theta, t) \right) \tag{6}$$

where the gap coordinates are given by θ and y as shown in Fig. 2. The pressure depends on the lubricant's viscosity given by μ and the clearance geometry given by $h(\theta, t)$.

The analytical solution of Eq. (6) regarding following boundary conditions

$$p(\theta, 0, t) = 0, \ p(\theta, B, t) = 0 \tag{7}$$

yields the pressure in a short bearing

$$p(\theta, y, t) = \frac{\mu}{h^3(\theta, t)} \left(12 \frac{\partial}{\partial t} h(\theta, t) + 6\dot{\varphi}_R(t) \frac{\partial}{\partial \theta} h(\theta, t) \right) \underbrace{\frac{y^2 - By}{2}}_{\leq 0} \tag{8}$$

whereas the explicit expressions of the gap height and its partial derivatives are given according to [13] by

$$h(\theta, t) = R_i - r + w(\theta) - x_R(t) \sin\theta - z_R(t) \cos\theta, \tag{9}$$

$$\frac{\partial}{\partial t} h(\theta, t) = \frac{\partial w}{\partial t} - \dot{x}_R(t) \sin\theta - \dot{z}_R(t) \cos\theta, \tag{10}$$

$$\frac{\partial}{\partial \theta} h(\theta, t) = \frac{\partial}{\partial \theta} w(\theta) - x_R(t) \cos\theta + z_R(t) \sin\theta. \tag{11}$$

As only quasi-static deformations of the bearing ring are considered, the partial derivative $\frac{\partial w}{\partial t}$ in Eq. (10) was omitted so far. Neglecting the velocity of the bearing ring reduces the effect of squeeze film damping and, hence, will be taken into account in future investigations.

After first integration with respect to y in Eq. (5) and consideration of Eqs. (9)–(11), the averaged pressure is given by

$$\bar{p}(\theta, t) = \mu B^3 \frac{\left(\dot{x}_R - \frac{\dot{\varphi}_R}{2} z_R \right) \sin\theta + \left(\dot{z}_R + \frac{\dot{\varphi}_R}{2} x_R \right) \cos\theta - \frac{\dot{\varphi}_R}{2} \frac{\partial w}{\partial \theta}}{(R_i - r + w - x_R \sin\theta - z_R \cos\theta)^3} \tag{12}$$

and with that the oil film force reads

$$\begin{pmatrix} F_{Bx}(t) \\ F_{Bz}(t) \end{pmatrix} = -\int_0^{2\pi} \overline{p}(\theta,t) \begin{pmatrix} \sin\theta \\ \cos\theta \end{pmatrix} r d\theta \qquad (13)$$

$$\approx -\sum_{m=0}^{N_\theta-1} \frac{1}{2} \begin{pmatrix} \overline{p}(\theta_m,t)\sin\theta_m + \overline{p}(\theta_{m+1},t)\sin\theta_{m+1} \\ \overline{p}(\theta_m,t)\cos\theta_m + \overline{p}(\theta_{m+1},t)\cos\theta_{m+1} \end{pmatrix} r\Delta\theta \qquad (14)$$

where the trapezoidal rule is used for the approximation of the second integration with respect to θ and according to Gümbel's boundary conditions only non-negative values $\overline{p}(\theta_m, t) \geq 0$ at $\theta_m = m\Delta\theta$ with $\Delta\theta = \frac{2\pi}{N_\theta}$ are added to account for cavitation phenomena.

2.2 Flexible Bearings

The elastically deformable outer bearing ring depicted in Fig. 2 is modeled as a thin and inextensible curved Euler-Bernoulli beam [16–18]. In the following, the small deformation due to an arbitrary number N_c of actuators is derived such that $0 < \theta_1 < \theta_2 < \cdots < \theta_{N_c} \leq 2\pi$ holds. The radial displacement w and tangential displacement u are coupled and both functions with respect to the coordinate θ.

Stress Resultants. The system of ordinary differential equations

$$\frac{d}{d\theta} N(\theta) + Q(\theta) + R_m n(\theta) = 0 \qquad (15)$$

$$\frac{d}{d\theta} Q(\theta) - N(\theta) + R_m q(\theta) = 0 \qquad (16)$$

$$\frac{d}{d\theta} M(\theta) + R_m \frac{d}{d\theta} N(\theta) + R_m^2 n(\theta) = 0 \qquad (17)$$

with the corresponding initial values

$$N(0) = N_0, \quad Q(0) = Q_0, \quad M(0) = M_0 \qquad (18)$$

describe the internal stresses N, Q and M of the curved Euler-Bernoulli beam in static equilibrium (see Fig. 3) where the radius of the middle fiber is R_m and the external loads are given by

$$n(\theta) = \frac{1}{R_m} \sum_{i=1}^{N_c} F_{ix} \delta(\Delta\theta_i) \qquad (19)$$

$$q(\theta) = \frac{1}{R_m} \sum_{i=1}^{N_c} F_{iz} \delta(\Delta\theta_i) \qquad (20)$$

in terms of unit impulses $\delta(\Delta\theta_i)$ at $\Delta\theta_i = \theta - \theta_i$. The external point loads

$$F_{ix} = -k_n u(\theta_i) \qquad (21)$$

$$F_{iz} = F_i - k_q w(\theta_i) \qquad (22)$$

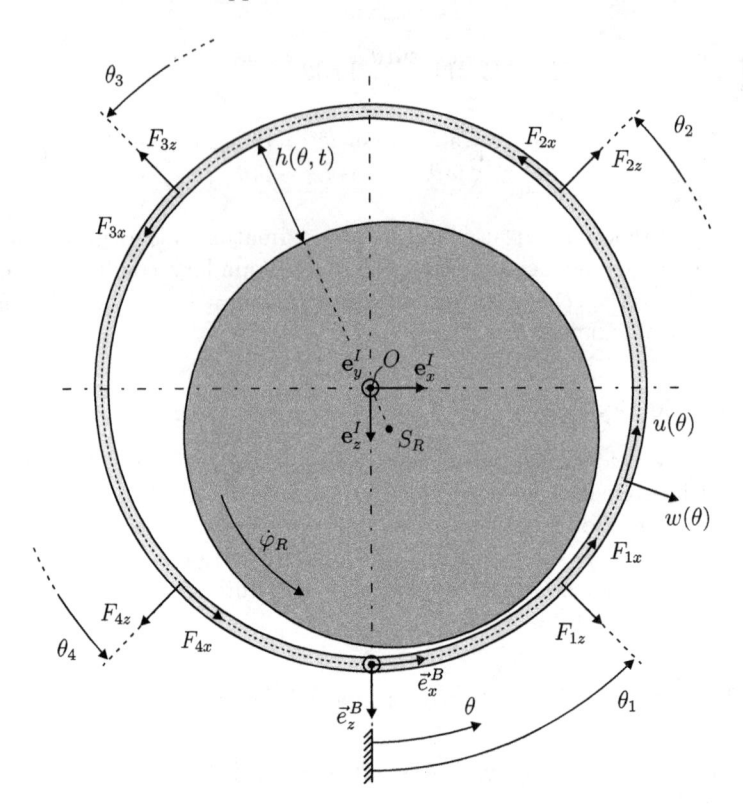

Fig. 2. Scheme of a flexible bearing with four evenly distributed actuators

For $i = 1, \ldots, N_c$ include both the control forces F_i of the actuators and the restoring forces due to the elasticity of the actuators. The stiffness in tangential and radial direction is given by k_n and k_q, respectively. In Eqs. (19) and (20), the fluid forces caused by the pressure and shear stress are neglected, i.e. the interaction between elastic bearing ring and viscous lubricant is assumed to be small. This assumption can be fulfilled if both the stiffnesses k_n, k_q of the ring support as well as the bending stiffness EI of the ring are sufficiently large and the eccentricity of the rotor remains in a moderate range. However, this fluid-structure interaction will be taken into account in future investigations, such that thinner and more compliant rings can be considered.

Solving the initial value problem given by Eqs (15)–(18) yields

$$N(\theta) = N_0 \cos\theta - Q_0 \sin\theta - \sum_{i=1}^{N_c}(F_{ix}\cos\Delta\theta_i - F_{iz}\sin\Delta\theta_i)\sigma(\Delta\theta_i) \quad (23)$$

$$Q(\theta) = N_0 \sin\theta + Q_0 \cos\theta - \sum_{i=1}^{N_c}(F_{ix}\sin\Delta\theta_i + F_{iz}\cos\Delta\theta_i)\sigma(\Delta\theta_i) \quad (24)$$

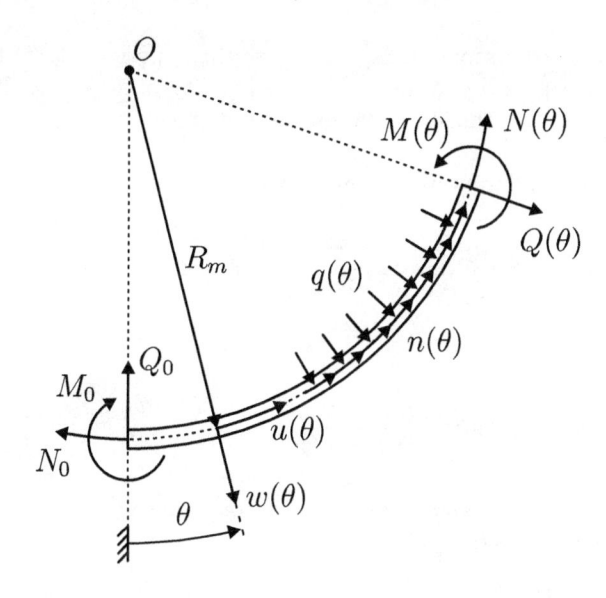

Fig. 3. Stress resultants within outer bearing ring

$$M(\theta) = M_0 + N_0 R_m (1 - \cos\theta) + Q_0 R_m \sin\theta \qquad (25)$$
$$-R_m \sum_{i=1}^{N_c}(F_{ix}(1-\cos\Delta\theta_i) + F_{iz}\sin\Delta\theta_i)\sigma(\Delta\theta_i)$$

in terms of the unit steps $\sigma(\Delta\theta_i)$ where $\sigma(0) := 1$ holds.

Deformations. The radial and tangential displacements of the thin, inextensible ring due to bending can be calculated by the ordinary differential equations

$$\frac{EI}{R_m^2}\left(\frac{\mathrm{d}^2}{\mathrm{d}\theta^2}w(\theta) + w(\theta)\right) = -M(\theta) \qquad (26)$$

$$\frac{\mathrm{d}}{\mathrm{d}\theta}u(\theta) + w(\theta) = 0 \qquad (27)$$

and corresponding boundary conditions

$$u(2\pi) = u(0),\ w(2\pi) = w(0),\ \frac{\mathrm{d}w}{\mathrm{d}\theta}(2\pi) = \frac{\mathrm{d}w}{\mathrm{d}\theta}(0), \qquad (28)$$

$$N(2\pi) = N(0),\ Q(2\pi) = Q(0),\ M(2\pi) = M(0). \qquad (29)$$

Due to the inextensibility assumption the radial displacement w and tangential displacement u are coupled by the kinematic constraint in Eq. (27). Eqs. (26) and (27) can be modified and extended in order to account for extensibility and shear.

The boundary conditions in Eq. (28) ensure the periodicity of the displacements, whereas Eq. (29) guarantees the periodicity of stress resultants on the one hand, and requires the static equilibrium of the external loads on the other hand.

Integrating the differential Eqs. given by Eqs. (26) and (27) yields

$$u(\theta) = C_1 + C_2 \cos\theta + C_3 \sin\theta + \frac{R_m^2}{4EI}\left(\tilde{C}_0(\theta) - 2R_m \sum_{i=1}^{N_c} \tilde{C}_i(\theta)\sigma(\Delta\theta_i)\right) \quad (30)$$

$$w(\theta) = -\frac{d}{d\theta}u(\theta) \quad (31)$$

$$= C_2 \sin\theta - C_3 \cos\theta - \frac{R_m^2}{4EI}\left(\frac{d\tilde{C}_0}{d\theta} - 2R_m \sum_{i=1}^{N_c} \frac{d\tilde{C}_i}{d\theta}\sigma(\Delta\theta_i)\right)$$

depending on the yet unknown integration constants C_1, C_2, C_3, N_0, Q_0 and M_0 and the coefficients $\tilde{C}_0$ and $\tilde{C}_i$ given by:

$$\tilde{C}_0(\theta) = 4(M_0 + R_m N_0)\theta + R_m(2N_0\theta - 3Q_0)\cos\theta - R_m(2Q_0\theta + 3N_0)\sin\theta,$$
$$\tilde{C}_i(\theta) = 2(F_{iz} + F_{ix}\Delta\theta_i) + (F_{ix}\Delta\theta_i - 2F_{iz})\cos\Delta\theta_i + (3F_{ix} + F_{iz}\Delta\theta_i)\sin\Delta\theta_i.$$

Using Eq. (28) the unknowns N_0, Q_0 and M_0 can be rewritten in terms of external point loads F_{ix} and F_{iz}. Then, substituting N_0, Q_0 and M_0 in Eqs. (21) and (22) the remaining unknowns C_1, C_2, C_3 can be determined by means of Eq. (29).

In the following, solutions of the above mentioned Eqs. are given for $N_c = 2$ with $\theta_1 = \pi$, $\theta_2 = 2\pi$ as shown in Fig. 1 such that $0 < \theta_1 < \theta_2 \leq 2\pi$ holds. Using the boundary conditions from Eq. (28)

$$u(2\pi) = u(0), \quad w(2\pi) = w(0), \quad \left.\frac{d}{d\theta}w(\theta)\right|_{\theta=2\pi} = \left.\frac{d}{d\theta}w(\theta)\right|_{\theta=0} \quad (32)$$

yields the explicit expressions

$$0 = \pi R_m\left(N_0 - \frac{F_{1x}}{6}\right) - \frac{2R_m F_{1z}}{3} + \frac{2\pi M_0}{3}, \quad (33)$$

$$0 = 2F_{1x} + \pi\left(Q_0 + \frac{F_{1z}}{2}\right), \quad (34)$$

$$0 = 2N_0 + F_{1x} \quad (35)$$

which can be solved for the unknowns

$$N_0 = -\frac{F_{1x}}{2}, \quad Q_0 = -\frac{4F_{1x} + \pi F_{1z}}{2\pi}, \quad M_0 = \frac{\pi R_m F_{1x} + R_m F_{1z}}{\pi}. \quad (36)$$

The boundary conditions from Eq. (29)

$$N(2\pi) = N(0), \quad Q(2\pi) = Q(0), \quad M(2\pi) = M(0) \quad (37)$$

imply the static equilibrium of the external loads:

$$0 = -F_{1x} + F_{2x}, \tag{38}$$
$$0 = -F_{1z} + F_{2z}, \tag{39}$$
$$0 = 2R_m F_{1x}. \tag{40}$$

The Eqs. (38)–(40) can be used together with Eqs. (21) and (22)

$$F_{1x} = -k_n u(\theta_1) \tag{41}$$
$$F_{1z} = F_1 - k_q w(\theta_1) \tag{42}$$
$$F_{2x} = -k_n u(\theta_2) \tag{43}$$
$$F_{2z} = F_2 - k_q w(\theta_2) \tag{44}$$

leading to the solutions

$$F_{1x} = F_{2x} = 0, \quad F_{1z} = F_{2z} = \frac{4\pi EI(F_1 + F_2)}{(\pi^2 - 8)R_m^3 k_q + 8\pi EI}, \tag{45}$$

$$C_1 = -4C_2 = -\frac{2\pi R_m^3 (F_1 + F_2)}{(\pi^2 - 8)R_m^3 k_q + 8\pi EI}, \tag{46}$$

$$C_3 = \frac{4\pi EI(F_1 - F_2) - R_m^3 k_q (4F_1 - 4F_2 + \pi^2 F_2)}{(\pi^2 - 8)R_m^3 k_q^2 + 8\pi EI k_q}. \tag{47}$$

For given integration constants the deformation of the outer bearing ring is described by Eqs. (30), (31) and can be used for pressure calculation as stated in Eq. (12).

2.3 Active Vibration Control

In general, the control forces F_1 and F_2

$$F_1 = F_{10} + F_1(z_R), \tag{48}$$
$$F_2 = F_{20} + F_2(z_R) \tag{49}$$

can be devided into static control forces (preload forces) F_{10} and F_{20} as well as dynamic control forces (feedback forces) $F_1(z_R)$ and $F_2(z_R)$, which model the displacement-proportional feedback control of an AVC system. According to the control forces, rigid body movements and deformations of the bearing ring can be realized simultaneously. It is well-known that non-circular bearings are able to suppress the onset of instability to a certain extent. In the context of active flexible bearings, this is for instance achievable by constant control forces $F_1 = F_{10}$ and $F_2 = F_{20}$ leading to a time-invariant non-circular bearing ring. Another possibility in this context is preloading the bearing ring first by static control forces $F_{10}, F_{20} \neq 0$ and regulating the deformation by dynamic control forces $F_1 = F_{10} + F_1(z_R)$ and $F_2 = F_{20} + F_2(z_R)$ when instability occurs. The preloading of the bearing ring is not considered in this investigation. Therefore, only the influence of dynamic control forces $F_1(z_R)$ and $F_2(z_R)$ is considered, while the static control forces F_{10} and F_{20} are set to zero.

In the following, the active flexible bearing as shown in Fig. 1 will be examined with a displacement-proportional feedback control with proportional gain K_P and linear as well as nonlinear actuator characteristics (see Fig. 4).

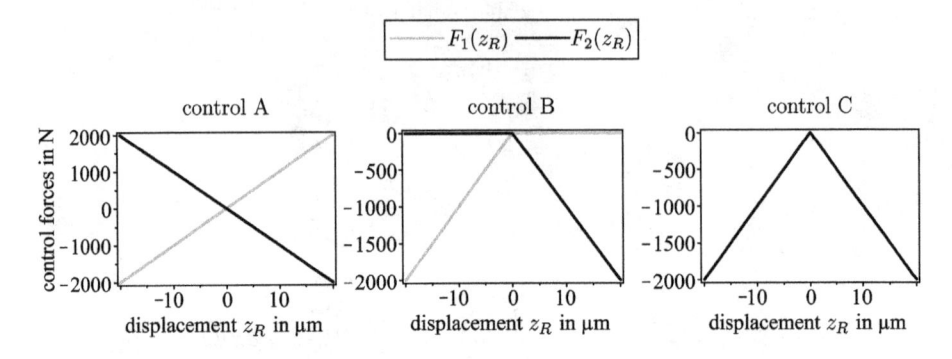

Fig. 4. Linear (A) and nonlinear (B,C) characteristics of the actuators

In the case of control A, only the rigid body movement is realized in vertical direction. Both actuators are simultaneously active and the control forces are given as

$$F_1(z_R(t)) = K_P z_R(t), \tag{50}$$
$$F_2(z_R(t)) = -K_P z_R(t). \tag{51}$$

For control B, only one of the actuators is active at a time while the other is in its dead zone,

$$F_1(z_R(t)) = \begin{cases} 0, & z_R(t) > 0 \\ K_P z_R(t), & z_R(t) \leq 0 \end{cases}, \tag{52}$$

$$F_2(z_R(t)) = \begin{cases} 0, & z_R(t) < 0 \\ -K_P z_R(t), & z_R(t) \geq 0 \end{cases}, \tag{53}$$

resulting in a two-lobed shape of the bearing ring, while its center of mass, however, moves in the vertical direction.

For control C, both actuators are active at a time and do not have dead zones,

$$F_1(z_R(t)) = -K_P |z_R(t)|, \tag{54}$$
$$F_2(z_R(t)) = -K_P |z_R(t)|, \tag{55}$$

that is, only a two-lobed shape is realized without movement of the center of mass. These three cases are be examined in detail and their potentials of vibration attenuation are demonstrated and compared to each other.

3 Vibrations of the Vertical Rotor System with AFBs

Run-ups of the vertical balanced rotor system on AFBs for each linear and non-linear actuator characteristics are simulated and compared to the uncontrolled rotor system. The parameters and initial conditions of the simulations are listed in Table 1 in Appendix C. The run-up simulation of the uncontrolled rotor system, i.e. $F_1 = F_2 = 0$, shown in Figs. 5 and 6 reveals unstable vibrations of the rotor system in the proximity of the equilibrium point $(x_R, z_R) = (0, 0)$.

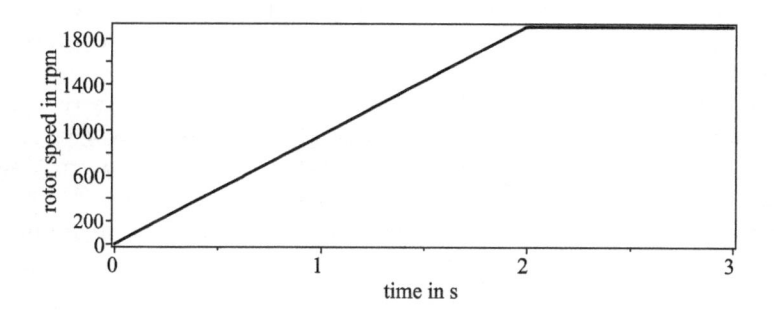

Fig. 5. Rotor speed during run-up

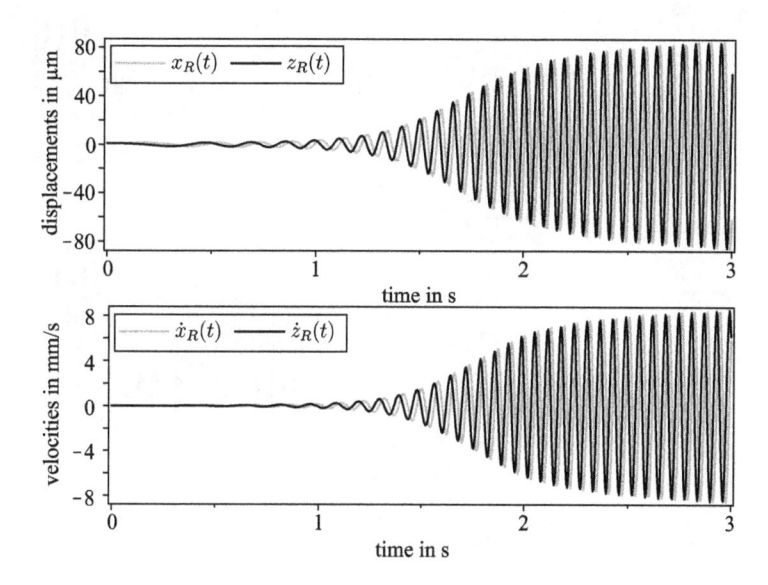

Fig. 6. Vibrations of uncontrolled rotor system

All three controls show a significant reduction in vibration but none of them is able to suppress the instability. Of the actuator characteristics shown in Fig. 4, control C achieved the best results in terms of vibration suppression (see Figs. 7, 9 and 11). For control C a reduction in vibration in both vertical and horizontal directions is achieved as shown in Fig. 7. The displacements of the uncontrolled rotor system are reduced from 80 μm to 0.5 μm, while the velocities are reduced from 8 mm/s to 0.05 mm/s. For control B the reduction in vibration is more significant in vertical direction than in horizontal direction as shown in Fig. 11. The horizontal displacements are reduced up to 3 μm, while the horizontal velocities are reduced to 0.5 mm/s. Similar results are achieved for control A in Fig. 9. The horizontal displacements are reduced up to 15 μm, while the horizontal velocities are only reduced to 7 mm/s.

In comparison, control B and C required lower control forces than control A. This means that the energy consumption of control A is higher due to the fact that for rigid body movements the actuators simultaneously work against the relatively stiff support of the bearing ring (see Figs. 8, 10 and 12). Control A requires control forces up to 300 N, while control B and C only require control forces up to 40 N.

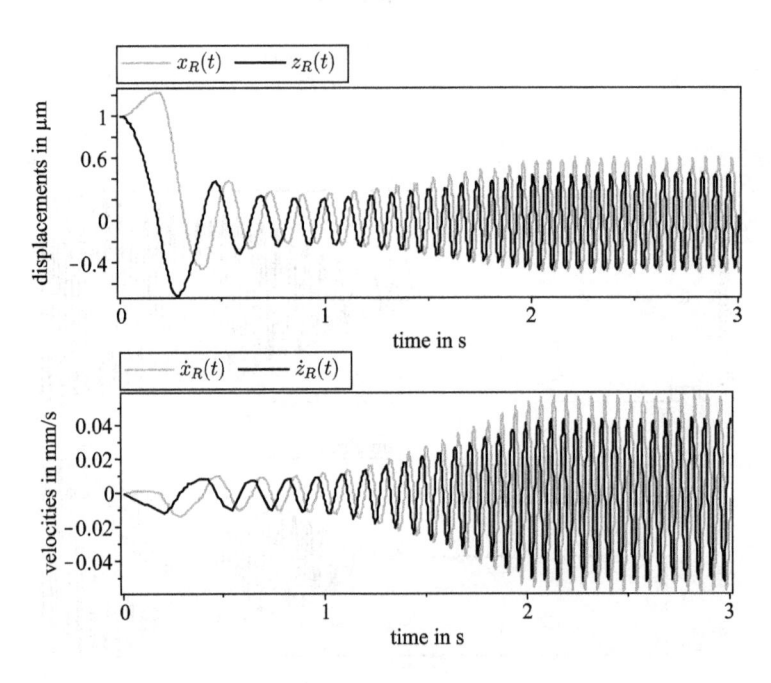

Fig. 7. Vibrations of the controlled rotor system with control C

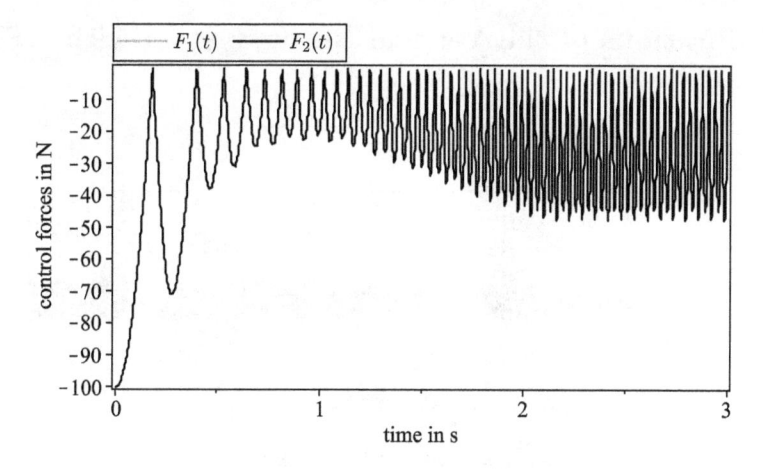

Fig. 8. Control forces of control C

4 Conclusions

A simple vertical rotor system with active flexible bearings is presented. The analytical solution of the deformations of the flexible bearing subject to point loads is derived. A displacement-proportional feedback control with both linear as well as nonlinear actuator characteristics is examined.

In the context of this active flexible bearing, actuator characteristics determine the combination of rigid body movements and deformations of the flexible bearing. Based on the presented numerical results rigid body movements of the flexible bearing require a higher energy consumption and should therefore be avoided. The two-lobed deformation that occurs synchronously with the rotor displacement show the most significant reduction in both vertical and horizontal direction. Nevertheless, all controls show a reduction in vibration but none of them is able to suppress the instability completely.

A Vibrations of the Vertical Rotor System with AFBs: Control A

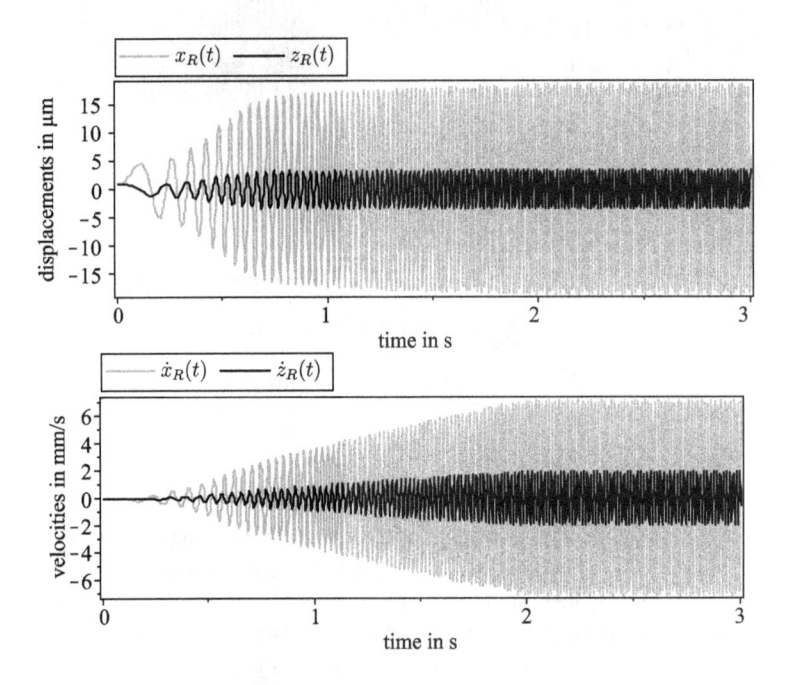

Fig. 9. Vibrations of the controlled rotor system with control A

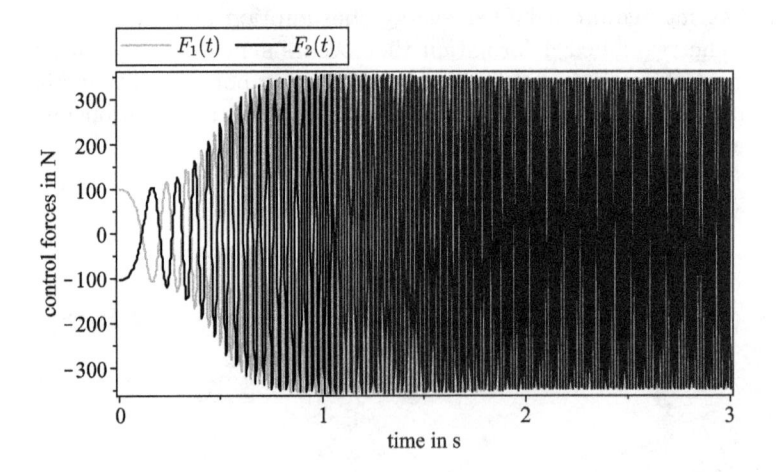

Fig. 10. Control forces of control A

B Vibrations of the Vertical Rotor System with AFBs: Control B

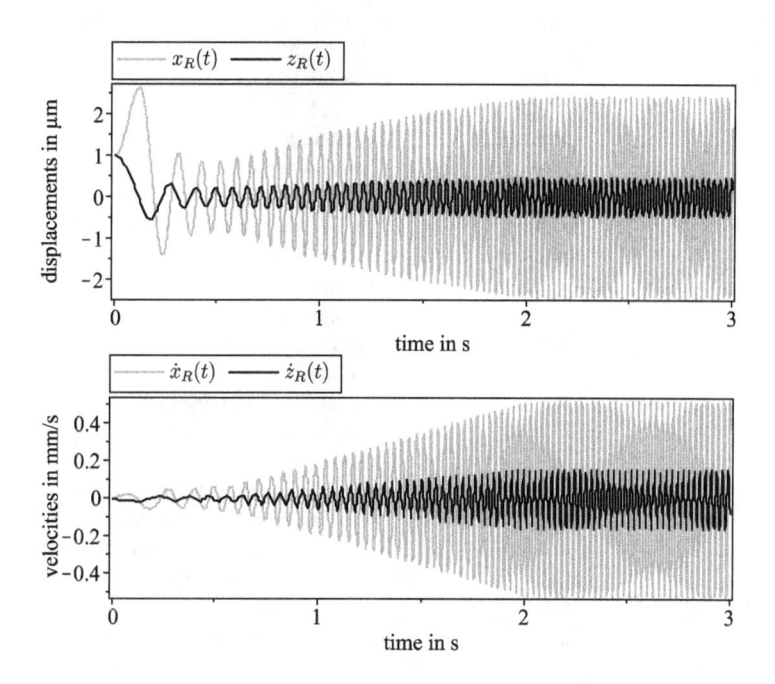

Fig. 11. Vibrations of the controlled rotor system with control B

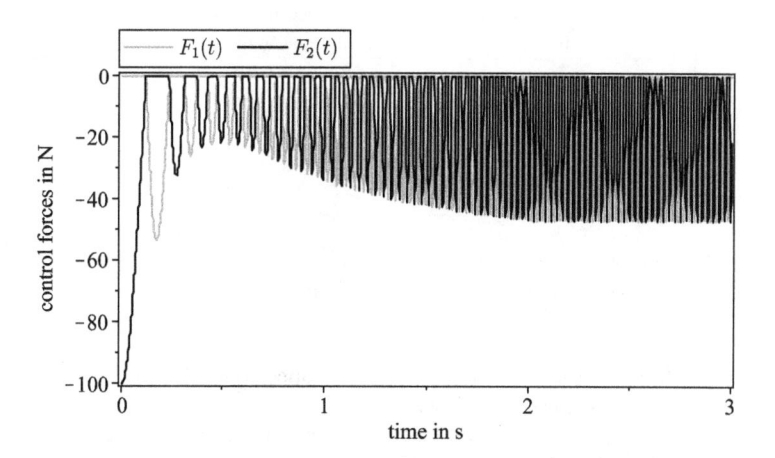

Fig. 12. Control forces of control B

C Parameters of the Vertical Rotor System with AFBs

Table 1. Parameters of the vertical rotor system with AFBs

Parameter	Value	Description
r	15 mm	shaft radius
ℓ	100 mm	shaft length
ρ	2700 kg kg/m^3	shaft density (steel)
$M_R = \rho \pi r^2 \ell$	0.190 kg	rotor mass
$J_R = M_R r^2/2$	$21.47 \cdot 10^{-6}$ kgm^2	rotor mass moment of inertia
M_{A0}	0.002 Nm	driving torque
H	1 mm	bearing ring thickness
B	5 mm	bearing length
μ	0.027 Pas	viscosity
c	100 µm	clearance
$R_i = r + c$	15.1 mm	inner radius of the bearing ring
$R_m = R_i + H/2$	15.6 mm	mean radius of the bearing ring
k_n, k_q	5 N/µm	support stiffnesses
E	420 MPa	Young's modulus of the bearing ring (PTFE)
$I = BH^3/12$	$4.166 \cdot 10^{-13}$ m^4	second moment of area of the bearing ring
K_P	100 N/µm	proportional gain of the feedback control
N_θ	75	number of intervals for trapezoidal rule
Δt	2 s	driving torque period
$\varphi_R(0)$	0	initial position of the rotor angle
$\dot\varphi_R(0)$	0	initial angular velocity of the rotor
$x_R(0)$	10^{-6} m	initial position of the rotor
$\dot x_R(0)$	0	initial velocity of the rotor
$z_R(0)$	10^{-6} m	initial position of the rotor
$\dot z_R(0)$	0	initial velocity of the rotor

References

1. Ishida, Y., Yamamoto, T.: Linear And Nonlinear Rotordynamics: a modern treatment with applications. Wiley, Hoboken (2013)
2. Kumar, A., Sinhasan, R., Singh, D.V.: Performance characteristics of two-lobe hydrodynamic journal bearings (1980)
3. Malik, M.: A comparative study of some two-lobed journal bearing configurations. ASLE Trans. **26**(1), 118–124 (1983)
4. Mehta, N.P., Singh, A.: Stability analysis of finite offset-halves pressure dam bearing. ASME. J. Tribol. **108**(2), 270–274 (1986). https://doi.org/10.1115/1.3261175
5. Dimond, T., Younan, A., Allaire, P.: A review of tilting pad bearing theory. Int. J. Rotating Mach. **2011**, 1–23 (2011)
6. Horst, H.G., Wölfel, H.P.: Active vibration control of a high speed rotor using PZT patches on the shaft surface. J. Intell. Mater. Syst. Struct. **15**(9–10), 721–728 (2004)

7. Krodkiewski, J.M., Cen, Y., Sun, L.: Improvement of stability of rotor system by introducing a hydraulic damper into an active journal bearing. Int. J. Rotating Mach. **3**(1), 45–52 (1997)
8. Krodkiewski, J.M., Sun, L.: Modelling of multi-bearing rotor systems incorporating an active journal bearing. J. Sound Vib. **210**(2), 215–229 (1998)
9. Sun, L., Krodkiewski, J.M.: Experimental investigation of dynamic properties of an active journal bearing. J. Sound Vib. **230**(5), 1103–1117 (2000)
10. Carmignani, C., Forte, P., Rustighi, E.: Active control of rotor vibrations by means of piezoelectric actuators. In: International Design Engineering Technical Conferences and Computers and Information in Engineering Conference, vol. 80272, pp. 757–764. American Society of Mechanical Engineers (2001)
11. Przybylowicz, P.M.: Active stabilisation of a rigid rotor by piezoelectrically controlled mobile journal bearing system. Aust. J. Mech. Eng. **1**(2), 123–128 (2004)
12. Tuma, J., Šimek, J., Škuta, J., Klecka, R.: The influence of controlled bushing movement on behaviour of a rotor in sliding bearings. Eng. Mech. **17**(3/4), 269–279 (2010)
13. Becker, K.U.: Dynamisches Verhalten hydrodynamisch gelagerter Rotoren unter Berücksichtigung veränderlicher Lagergeometrien (Doctoral dissertation, Karlsruher Institut für Technologie (KIT)) (2020)
14. Bäuerle, S.: An Approach to Non-linear Dynamics of Rotors with flexible Seals: Models. Kassel University Press, Numerical Tools and Basic Phenomena (2021)
15. Krämer, E.: Dynamics of Rotors and Foundations. Springer, Cham (2013)
16. Bitner, A., Seemann, W.: Deformations of a piezo-actuated bearing ring in journal bearings. PAMM **21**(1), e202100221 (2021)
17. Prescott, J.: Applied elasticity. Longmans, Green and Company (1924)
18. Flügge, W.: Statik und Dynamik der Schalen. Springer, Heidelberg (1934). https://doi.org/10.1007/978-3-642-52728-9

A Novel Monolithic Shape-Morphing Bearing for Real-Time NVH Control

Christos Kalligeros[✉], Georgios Chantoumakos, Efstratios Tsolakis[iD], Panagiotis Spiridakos, and Vasilios Spitas

Laboratory of Machine Design, National Technical University of Athens, 9 Heroon Polytechniou Str., 15780 Zografou, Greece
ckalligeros@mail.ntua.gr

Abstract. In this paper a novel, active, hydraulically actuated, shape-morphing 3-wave bearing is presented for the active control of noise and vibrations of rotating machines. Wave bearings are used in various critical mechanical structures since they offer better stability, higher load capacity and improved performance compared to classical journal bearings and in addition, due to their geometry, they can attenuate vibrations and reduce noise emissions by more than 20%. The novelty of the presented bearing design lies in its capability of adapting its geometry in real-time using hydraulic actuation through a monolithic simple pocket design. The design was validated using computational fluid dynamics (CFD) and fluid-structure interaction (FSI) analysis and a full dynamic simulation was performed to investigate the vibration control capabilities of the bearing as part of an operating powertrain. It is demonstrated that its shape-morphing characteristics may reduce vibration transfer up to 100%, while for low-stiffness systems anti-resonance responses can be achieved. Meanwhile, the design of the bearing is very simple, requiring minimum interventions in existing applications.

Keywords: Wave bearing · 3-wave bearing · shape-morphing · journal bearing · NVH performance

1 Introduction

Plain journal bearings are a well-known and established bearing solution for a wide range of engineering applications since they combine a compact and simple construction with considerable load capacity, great dampening ability of shaft oscillations thus resulting in lower noise emissions, high lifespan and high-speed operation. However they tend to suffer from instability problems especially in low-load conditions. To overcome this problem Dimofte [1–3] proposed in the early 90s a novel design of a journal bearing where the inner bearing diameter is not circular but has a wave-like profile. The so-called wave journal bearing offer larger load capacity than plain bearings and absorb to a greater extent the oscillations and vibrations of the shaft. Since today noise reduction is achieved through additional costly constructions that add weight and complexity to systems, wave bearings can offer a simple and effective solution to control noise and vibration levels to a variety of applications such as gas turbines, helicopter transmissions, pumps and machine tools [2, 4–7].

The improved stiffness and stability characteristics of the wave bearing compared to other types of journal bearings (mainly the plain bearing but also the groove or the three-lobe bearing) have been identified already by Dimofte in his first papers on wave bearings [1–4]. He also found that the wave bearing performance is dependent on the waves amplitude and position relative to the applied load. Yang et al. [7] studied the static behavior of wave bearings and concluded that their performance is dependent on all the main design parameters including eccentricity ratio, wave amplitude ratio, wave position angle and wave number. In most cases a wave number of 3 was found to produce the best load and stiffness characteristics. The design parameters of wave bearings influence also their dynamic performance being able to control the stability margins of the bearings [8–10].

The noise and vibration characteristics of wave bearings do not only surpass those of other types of journal bearings but also prevail over those of rolling bearings in most applications. Ene & Dimofte [11] incorporated wave bearings to attenuate the gear mesh noise produced in a helicopter transmission. It was found that at certain speeds and loads the wave bearings can offer a noise reduction by more than 10 dB compared to rolling elements. But even when not operating at the preferred speeds and loads experimental data showed that wave bearings still have comparable vibration characteristics with rolling bearings without the need for an additional vibration dampening system (as is often the case for rolling bearings) [6].

After observing the dependence of the dynamic characteristics of wave bearings to their design parameters, Dimofte highlighted already in 1993 [1] the possibility of a real-time control of the vibration response of a rotor-bearing system through the active adjustment of the wave amplitude of the bearings. The implementation of journal bearings with adjustable geometry/parameters can significantly improve the performance of rotating dynamic systems especially when operating at a wide range of speeds and loads [12]. The active control of the journal bearing geometry will be able to modify the stiffness and damping characteristics of the system introducing anti-resonances and leading to stable operation and noise attenuation [13].

The goal of this paper is to introduce a novel design for an active, hydraulically actuated, shape-morphing 3-wave bearing for the active control of noise and vibrations of rotating machines. Through a monolithic design and a carefully-executed pocket-placement a plain journal bearing can be converted into a wave bearing (or generally into any type of journal bearing without a circular inner diameter) in real-time or vice versa. The proposed design is simple and requires minimum innervations to existing rotating systems in order to be incorporated. At the same time, it is demonstrated through CFD, FSI and dynamic analysis that the shape-morphing wave bearing can contribute to a noise reduction of 100% at certain applications, while it can lead to an anti-resonance response of the system when required.

2 The Wave Journal Bearing

Wave journal bearings were created in the 1990s by Dimofte [1–3] in collaboration with NASA as an alternative type of journal bearing where the inner diameter is not circular but has a wave-like profile (Fig. 1). Compared to circular plain bearings they have some very interesting advantages due to this geometry such as increased load capacity, improved stability and greater absorption of shaft oscillations/vibrations to achieve significant noise reduction (the oscillations are damped in the wave bearings and are not transferred to the outer housing).

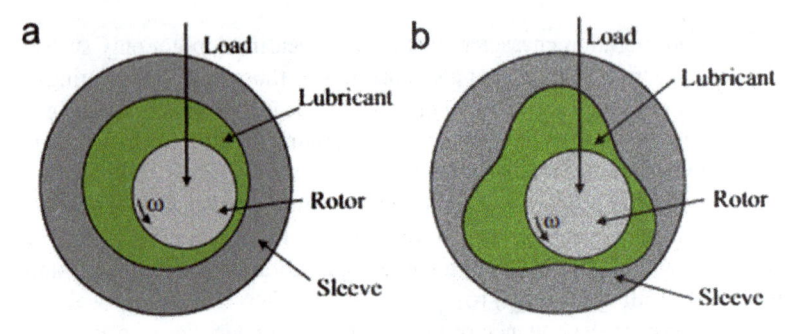

Fig. 1. (a) Plain journal bearing, (b) Wave journal bearing [11]

The first basic parameter of the wave bearing is the dimensionless wave amplitude ratio $\mathcal{E}_w$ which is expressed by the equation:

$$\varepsilon_w = \frac{e_w}{C} \tag{1}$$

where $\mathcal{E}_w$ is the amplitude of the wave and C is the radial clearance, but unlike in plain bearings it is measured as

$$C = R_m - R_s \tag{2}$$

where R_s is the radius of the shaft and R_m is the radius of the inner side of a hypothetical bearing that would pass through the centre of the wave as shown in Fig. 2.

The profile of the inner diameter of the wave bearing follows the equation:

$$y = R_m + e_w \sin(kx) \tag{3}$$

where k the number of the waves.

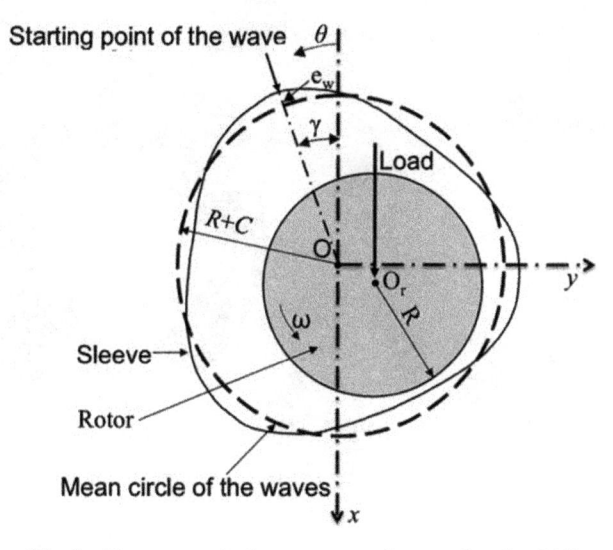

Fig. 2. The geometrical parameters of a wave bearing [11]

The lubricant thickness at every angle θ is given by the equation [14]:

$$h(\theta) = C + x\cos\theta + y\sin\theta + e_w\cos(n_w(\theta + \gamma)) \qquad (4)$$

where n_w is the number of waves and γ is the starting angle of the wave as shown in Fig. 2.

In this paper a three-wave journal bearing is considered. In bearings with four, five or more waves, the effect of the presence of the waves is attenuated due to poor fluid flow development resulting in less advantageous characteristics [2, 3].

3 Concept of the Novel Shape-Morphing Journal Bearing

The basic concept behind the shape-morphing journal bearing is the application of external pressures to a cylindrical bearing in order to be deformed in a wave-like shape and perform as a wave bearing. This is possible because the wave amplitude of wave bearings is small in amplitude (in the order of a few microns) so a reasonable amount of external pressure can artificially create a waved geometry on the inner surface of the bearing. By actively modifying the external pressures acting on the bearing, the wave amplitude can be changed in real-time and thus an active control of the vibration response of the system can be achieved.

The main design of the shape-morphing journal bearing is presented in Fig. 3. The bearing constitutes of two rings one inside the other. At certain areas of its outside surface the inner ring is not completely cylindrical. In contrast, the outer ring is completely cylindrical and fastens on the outside side of the inner ring. At the areas where the inner ring is not cylindrical, a gap will form between the two rings. These points are the so-called pockets of the bearing where the external pressure will be hydraulically actuated to deform the inner surface of the bearing into a wave-like shape.

86   C. Kalligeros et al.

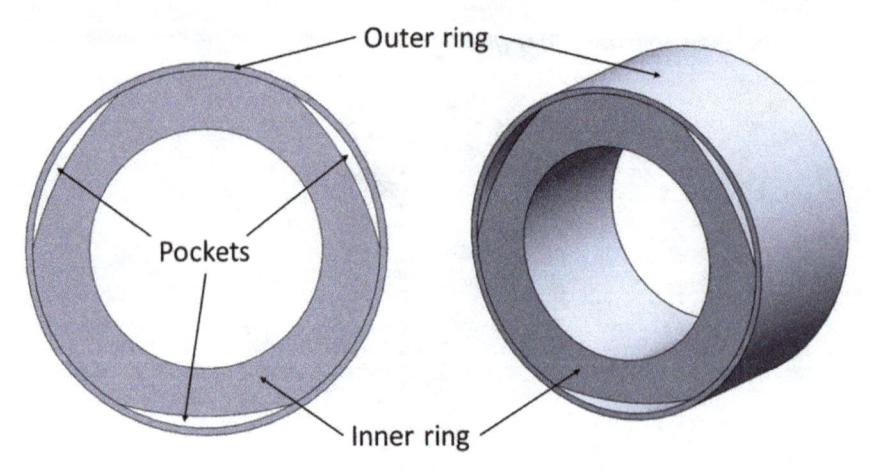

Fig. 3. Schematic representation of the novel shape-morphing journal bearing design

There are a lot of geometrical parameters affecting the deformation of the inner ring into a bearing with a waved inner surface. Figure 4 illustrates the geometry of the inner ring and its main parameters. In Fig. 4 D_b is defined as the diameter of the shaft and t_{b1} is defined as the thickness of the inner ring.

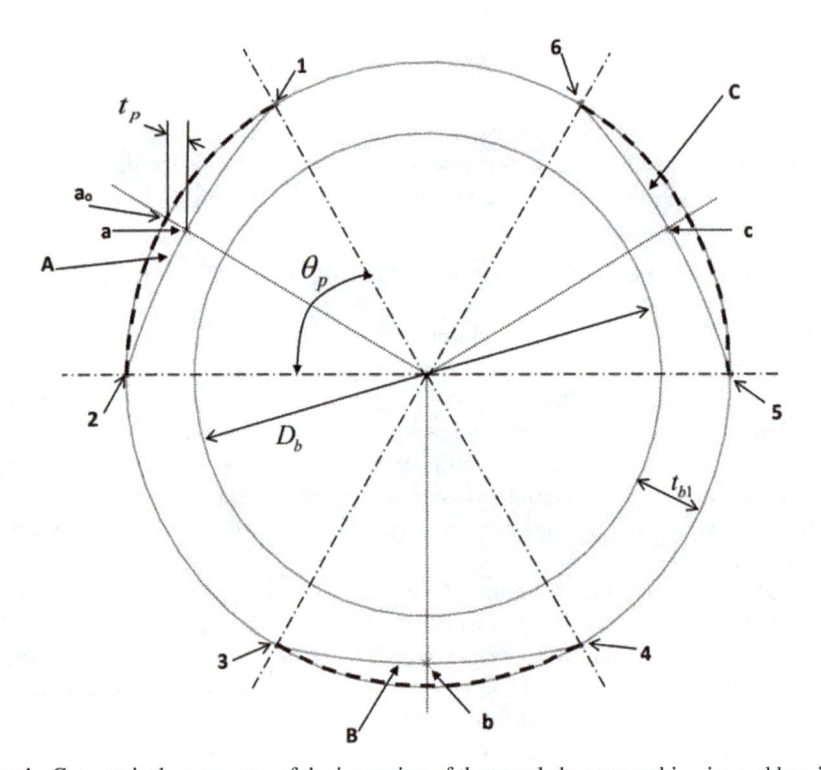

Fig. 4. Geometrical parameters of the inner ring of the novel shape-morphing journal bearing

Arcs A, B and C are located at the position of the three pockets of the bearing. Arc A begins at point 1 and ends at point 2, where a is the intermediate point of the arc. The dotted lines connecting 1 and 2 represent the outside surface of the inner ring if it was perfectly circular. The gap formed between point a and point a_0 (Fig. 4) is defined as the thickness t_p of the pocket. θ_p is the position angle of the arc A. The geometrical parameters for arcs B and C are defined in a similar manner.

The main factors affecting the magnitude of the external pressures that must be applied to the pockets to deform the cylindrical inner ring into a wavy one are:

- The thickness of the bearing t_{b1}
- The desired amplitude of the wavelength on the inner side of the bearing
- The magnitude of the pressures generated inside the bearing during operation and the positions where the pressures occur
- The thickness t_p of the pockets and the arcs A, B, C which will be defined based on the three previous factors.

In addition, the thickness of the pockets and in general the positioning of A, B, C is also influenced by manufacturing and operational factors such as the providence of sufficient space for the lubrication system and the assurance of the structural integrity of the bearing assembly.

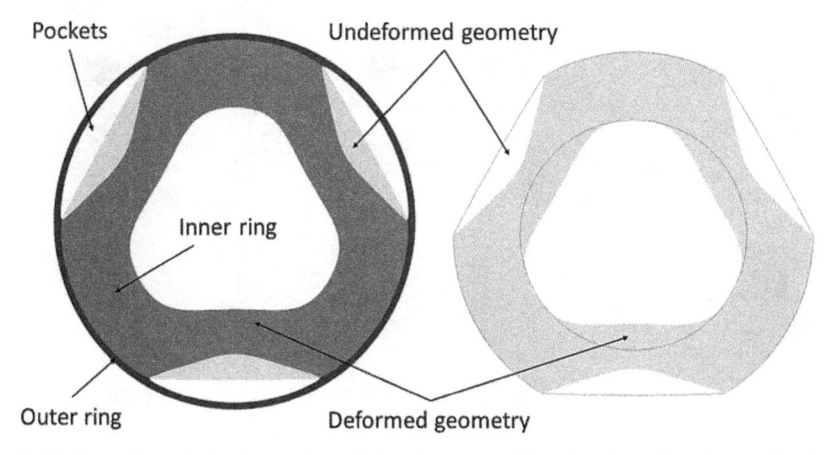

Fig. 5. Deformation of the inner ring of the shape-morphing bearing after the application of external pressure to the pockets. The transparent part illustrates the undeformed geometry, while the shaded part the deformed one. The outer ring is illustrated schematically.

Figure 5 illustrates the deformed bearing when external pressures are applied to the pockets. The transparent part shows the undeformed geometry of the inner ring while the shaded part shows the deformed inner ring after the application of the external forces in the pockets and the formation of a wave-like profile to the inner diameter. Despite this, it is difficult to deform the inner surface of the bearing in such a way as to accurately produce a wave bearing such as in Figs. 1 and 2. However, as will be shown on Sect. 4.4, there are no significant differences to the characteristics of a wave bearing whose inner surface does not match exactly (up to a certain extent) the nominal wave profile. Finally, the thickness of the outer ring should be calculated so the deformation of the outer ring is negligible (i.e. an order of magnitude lower) compared with the deformation of the inner ring.

4 Modeling

4.1 FEA Modeling

The simulation of the computational fluid dynamics (CFD) problems where performed through Ansys Fluent. Both the plain and the wave journal bearing are studied in order to have a clear reference for the performance of the latter.

For the modeling of the journal bearings the cylinder was divided into 360 segments, while three (3) finite elements were considered along the thickness as shown in Fig. 6. The produced mesh consisted of approximately 3 million elements and 4 million nodes. Hexagonal finite elements were used instead of tetrahedral finite elements to improve accuracy and computational time.

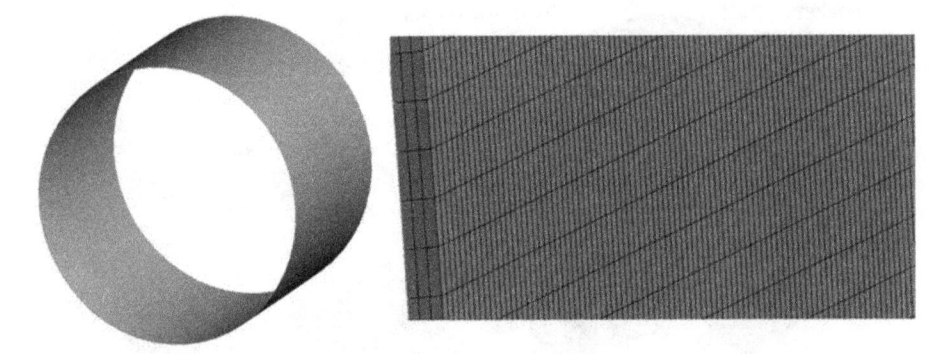

Fig. 6. Mesh for lubricant film of a journal bearing in ANSYS Fluent

The produced FEA model is benchmarked through published results in literature. Dhande et al. [15] performed a Fluid Structure Interaction (FSI) analysis to a plain journal bearing whose parameters are given in Table 1.

Table 1. Geometrical and lubricant properties of the benchmark plain journal bearing as per [15]

Parameter	Values
Shaft Radius (mm)	50
Length (mm)	80
Radial clearance (μm)	145
Rotational speed (rpm)	4000–10000
Viscosity (Pas)	0.0277
Density (kg/m^3)	860

The boundary conditions and the remaining parameters have been selected as follows:

- The flow is laminar
- The inner side of the bearing is constant
- The sides of the lubricant layer that are not adjacent to either the bearing or the shaft were set as the fluid inlet side and the fluid outlet side (common approach in CFD analyses).

The first comparison was for a short length bearing with an eccentricity ratio of $\varepsilon = 0.2$ and a shaft rotation speed of N = 4000 rpm. The pressures formed are similar and are shown in Fig. 7.

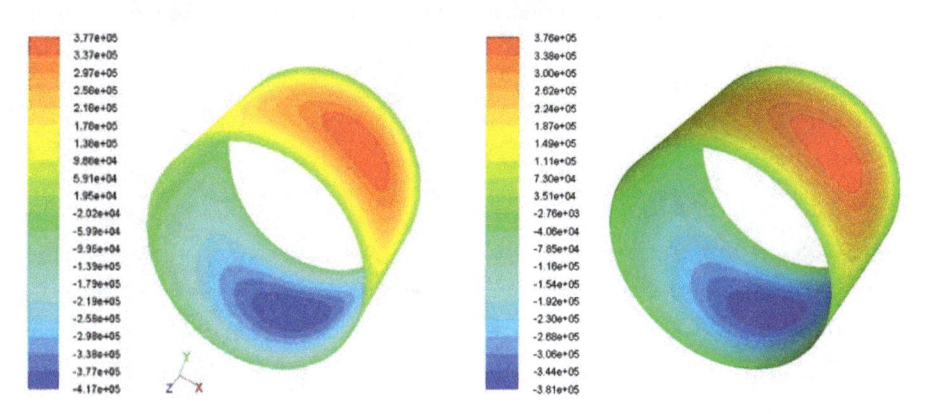

Fig. 7. Pressure distribution for $\varepsilon = 0.2$ as per [15] (left) and from the present model (right)

The second comparison refers to a constant rotational speed N = 4000 rpm and a variable eccentricity ratio e. The results are shown graphically in Fig. 8.

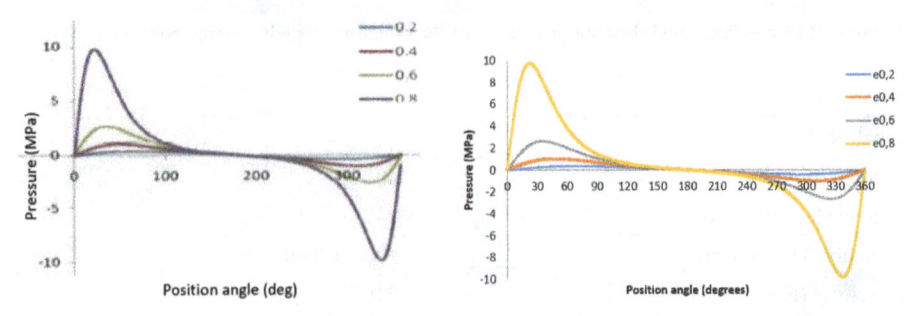

Fig. 8. Static pressure for different eccentricity ratios as per [15] (left) and from the present.

In all three comparisons made between the two models the pressures are essentially identical.

4.2 Comparison of Plain and Wave Journal Bearings

The model presented above is used to examine the behavior of the plain and wave journal bearings at low pressures in order to investigate Dimofte's [2] claim that at low eccentricities the wave bearing offers stability. When the shaft is in the center or close to the center (namely the eccentricity is close to zero) the dynamic response of the system mounted on plain bearings becomes unstable and there is a significant risk of damage. While in a plain bearing no pressures occur, a three-wave wave bearing develops three pressure maxima which provide sufficient stability and high stiffness to the system as shown in Fig. 9 for $\varepsilon = 0$.

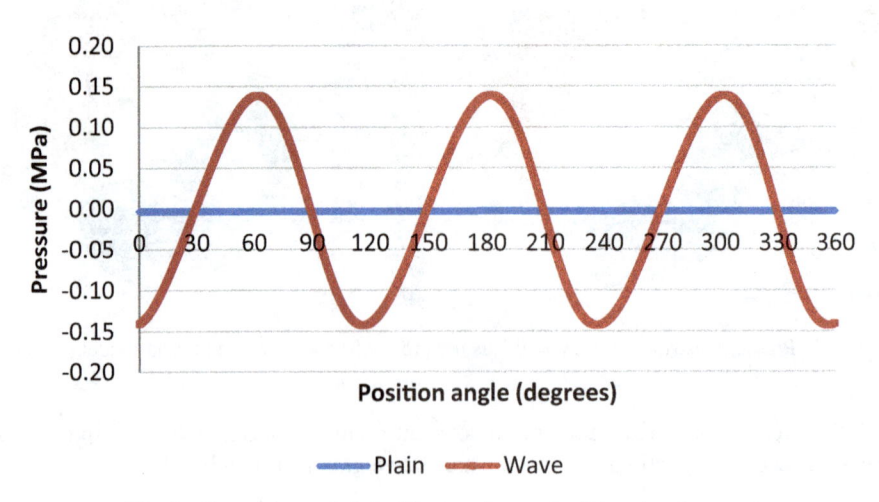

Fig. 9. Comparison of plain (blue) and wave (red) bearing for $\varepsilon = 0$

The above findings are very important because short length plain bearings, for small L/D ratios < 0.3, have poor damping in oscillations. This is one of the important problems that the wave bearing comes to solve.

Figure 10 presents the pressure distribution for constant rotational speed N = 4000 RPM and different eccentricity ratios. It is observed that for small eccentricity ratios the effect of waves on the wave bearings is significant. As the eccentricity ratio increases the effect of the waves is eclipsed by the maximum pressures (i.e. for $\varepsilon = 0.6$).

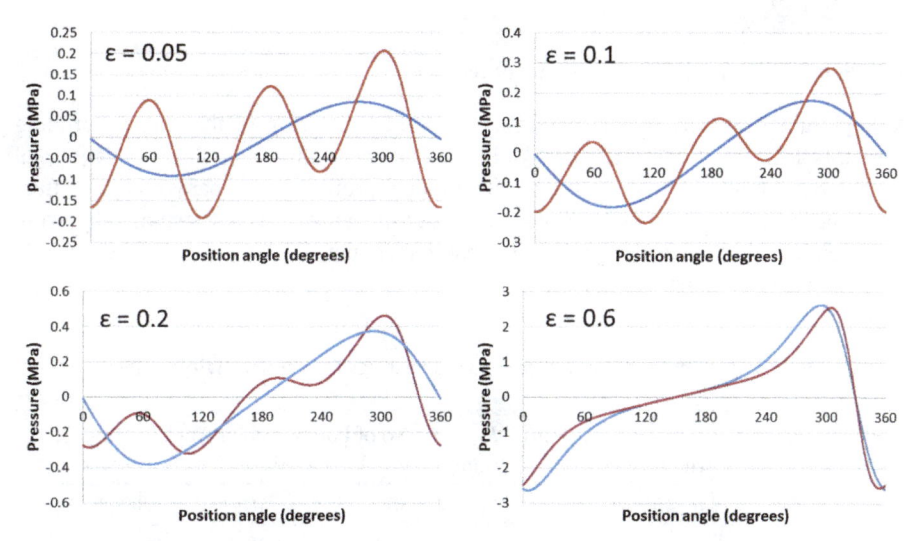

Fig. 10. Comparison of plain (blue) and wave (red) bearing for $\varepsilon = 0.05, 0.1, 0.2, 0.6$

Figure 11 presents the pressure distribution for eccentricity ratio $\varepsilon = 0.2$ and different rotation speeds. For low rotational speeds the effect of waves on the shaft bearing is less significant. As the rotational speeds increase, their effect becomes more pronounced.

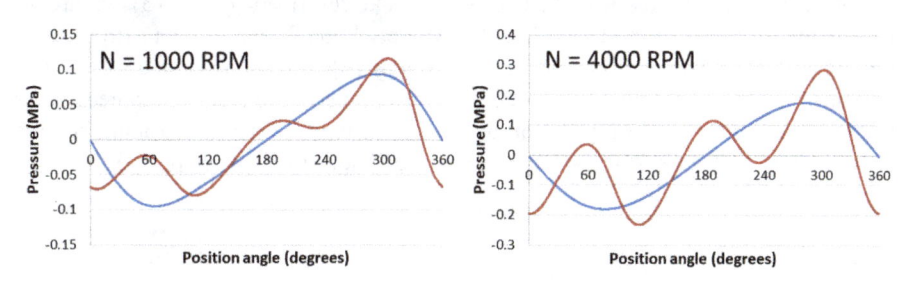

Fig. 11. Comparison of plain (blue) and wave (red) bearing for $\varepsilon = 0.2$ and N = 1000 RPM and 4000 RPM

The same trends apply to other values of eccentricities (i.e. $\varepsilon = 0.4$–0.8) and other rotational speeds (i.e. $N = 6000$–10000). The effect of waves is significantly stronger when the wave ratio ε_w is high. However, even in this case where the wave ratio is $\varepsilon_w \approx 0.1$ the geometry of the wave bearings have a positive effect on the bearing-axis stability and reduce the shaft oscillations.

The above findings are very important especially for short length sliding bearings (i.e. L/D ratios lower than 0.3) that demonstrate poor damping in oscillations. This is one of the important problems that the wave bearing comes to solve.

4.3 Sensitivity Analysis of Lobe Geometry

As highlighted in Sect. 3, the deformation of the inner ring of the shape-morphing journal bearing can lead to a wave-like inside diameter similar to a three-wave journal bearing but not necessarily identical to its nominal profile. In this paragraph the effect of geometrical deviations to the characteristics of the wave bearings is studied by using three bearings with the same geometrical parameters but slightly different inner profile were designed as shown in Table 2.

Table 2. Geometrical parameters of wave bearings for the sensitivity analysis

	R (mm)	C' (mm)	R_{min} (mm)	R_{max} (mm)	No of Points	Point Angle	Wave Start Point
Model 1	50	0.145	50.145	50.174	36	10°	90°
Model 2					360	1°	
Model 3					Continuous	Continuous	

Models 1 and 2 are designed by placing a point at every 10° and 1°. The points are then connected by linear segments. Model 3 represents a bearing with a continuous sinusoidal inner profile as described in Sect. 2 and described by Eq. (3). Figure 12 presents the geometry of models 1, 2 and 3. Models 1 and 2 represent an extreme case with significant geometrical deviations, much bigger than those expected to occur in the shape-morphing bearing. Consequently, if there are no significant differences to the characteristics of these bearings it can be concluded that the shape-morphing bearing can operate as a wave bearing even with minor deviations to the wave profile.

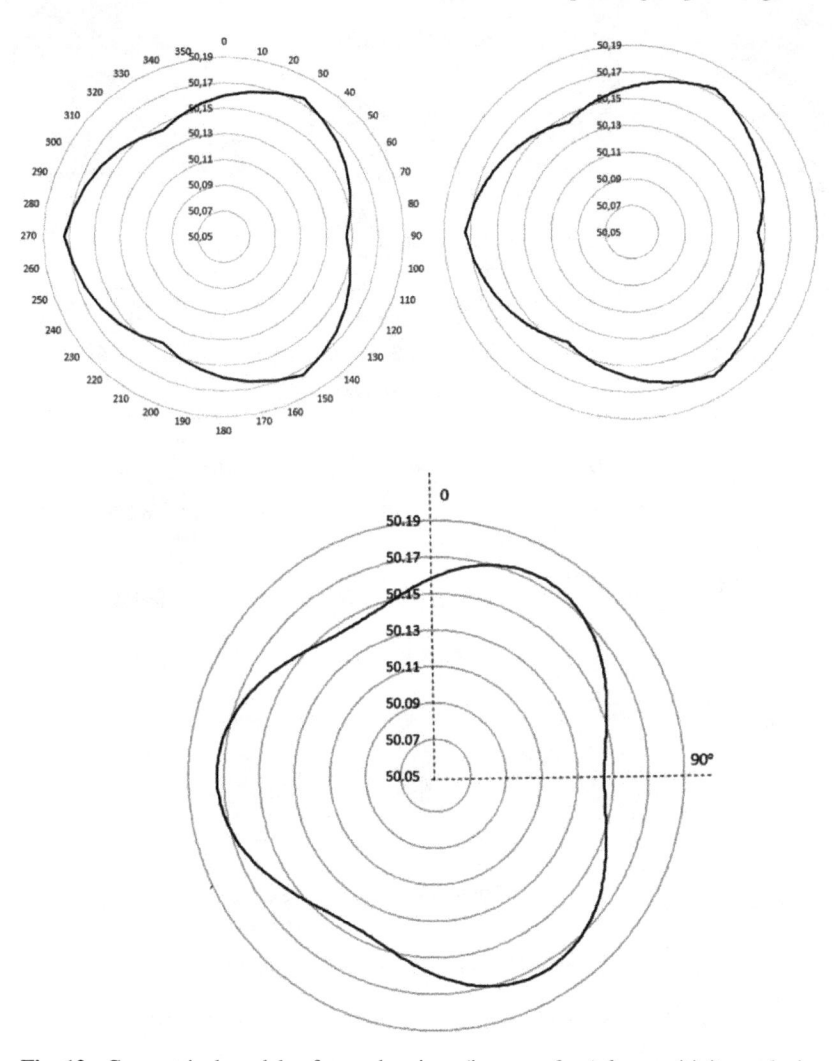

Fig. 12. Geometrical models of wave bearings (inner surface) for sensitivity analysis

Figures 13, 14 and 15 present the maximum pressure, the pressure distribution and the total force respectively at different points of the bearings. The eccentricity ratio is ε = 0.2 and the rotational speed is N = 4000 RPM.

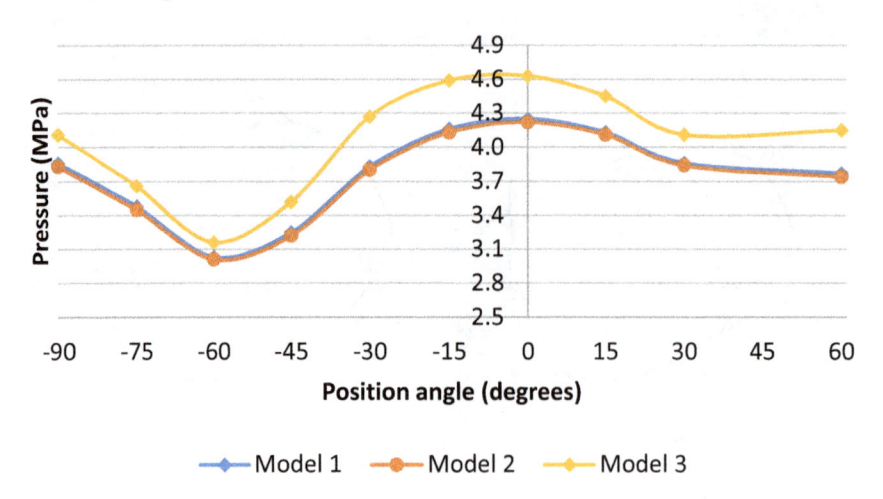

Fig. 13. Comparison of maximum pressure for models 1, 2 and 3 from 270° to 60°

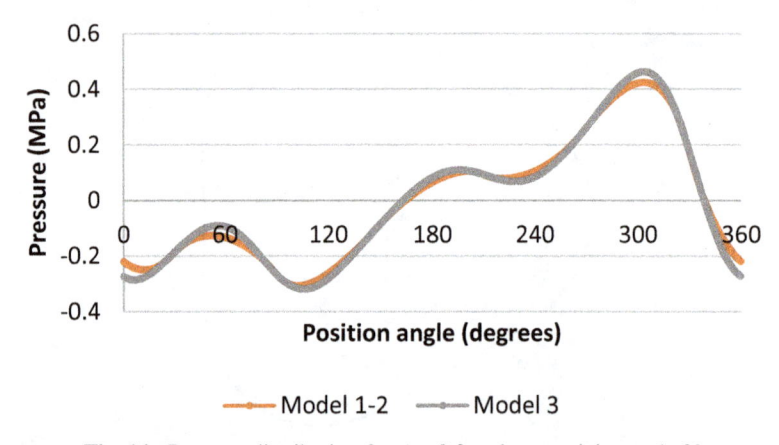

Fig. 14. Pressure distribution for $\varepsilon = 0.2$ and eccentricity angle 0°

Models 1 and 2 have almost the same maximum pressures, while there is a difference of 8.9% when compared to model 3. Despite that, the difference at the total forces is only 0.15%. Consequently, it can be concluded that even for these extreme cases the deviations of the inner profile of the wave bearing do not have a significant effect on its parameters. However, the above conclusion is not necessary valid if the endpoints R_{max} and R_{min} are changed or are not common. More work needs to be done in this area for the conclusions to be generalized to include that case.

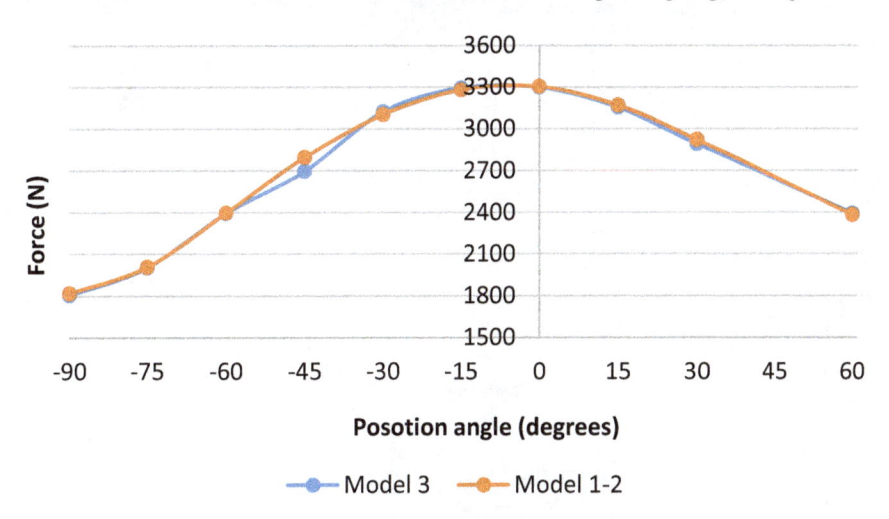

Fig. 15. Total force on bearing for models 1, 2 and 3

4.4 Boundary Conditions for Negative Pressures

The finite element analysis presented above predicted both positive and negative pressures for the lubricant film. However, negative pressures are impossible to occur in a real journal bearing during operation. The only negative pressures that can occur are when locally the lubricant pressure can decrease below the ambient pressure resulting in bubbles and the occurrence of cavitation. The region where negative pressures occur is usually considered as inactive and is not taken into account in CFD analyses [15, 16]. However, this assumption would not be valid for an FSI problem, since it would result to inaccurate stress and deformation fields on the bearing. For the wave journal bearing of Table 1 with an eccentricity ratio of $\varepsilon = 0.6$ and an eccentricity angle $-30°$ (or $330°$) the maximum positive and negative pressure was 2.55 MPa and -2.55 MPa respectively. The total force was calculated as 4.7×10^3 kN at an angle $\varphi = 90°$, namely at the direction of the external load, validating the equilibrium of forces. If the negative pressures were to be neglected (i.e. to be replaced by a zero value of pressure) the total force would be 3.1×10^3 kN at an angle $\varphi = -51°$ disrupting the force equilibrium of the bearing.

There are two possible ways to cope with this problem, the first is the adoption of an iterative procedure during which only negative pressures calculated at each step are zeroed and this condition is introduced as input for the next iterative step where all pressures are recalculated until the method converges to a result with non-negative values. This however produces solutions with displaced centers of pressure on the bearing and in order to avoid these the second alternative was followed according to which a uniform external positive pressure DP was superimposed on the pressure field calculated by FEA. To eclipse negative pressures a positive external pressure equal to the maximum negative pressure must be applied. As shown in Fig. 16 the pressures shift by almost 2.5 MPa (i.e. 25 bar).

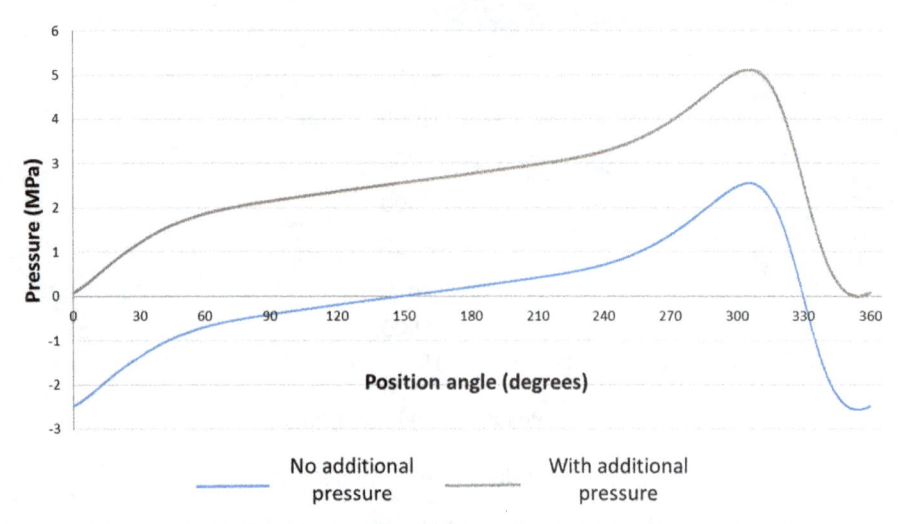

Fig. 16. Pressure distribution with (grey) and without (blue) additional external positive pressure

5 Design of the Novel Shape-Morphing Bearing

5.1 Main Design Parameters

The placement and size of the pockets of the shape-morphing bearing must meet certain requirements:

1. They must be positioned so as to create a wave profile similar to Eq. (3). Consequently the pockets at 90°, 210° and 330° will be designed in a way that when external pressures are applied the radius of the inner surface at these points becomes 50.145 mm.
2. In the same way, the radius of the inner surface at 30°, 150° and 270° should be 50.174 mm. Therefore in these points the deformation should be zero.

The detail design of the pockets to meet the aforementioned requirements was done with structural analysis in ANSYS Workbench. The main parameters for the design of the pockets was defined as:

- The eccentricity ratio $\varepsilon_w = \frac{e_w}{C}$, as in Eq. (1).
- The thickness ratio $t_w = \frac{t_p}{t_{b_1}}$, where t_p is the thickness of the pocket and t_{b_1} is the thickness of the inner ring.
- The angle ratio $\theta_w = \frac{\theta_p}{\theta_{ab}}$ where $\theta_p = \theta_2 - \theta_1$ is the central angle of arc A and θ_{ab} is the angle between 2 consecutive wave maxima, namely $\theta_{ab} = \frac{360}{k}$, as illustrated in Fig. 17. For a three-wave wave bearing $\theta_{ab} = 120°$. θ_w is calculated for arcs B and C in a similar way.

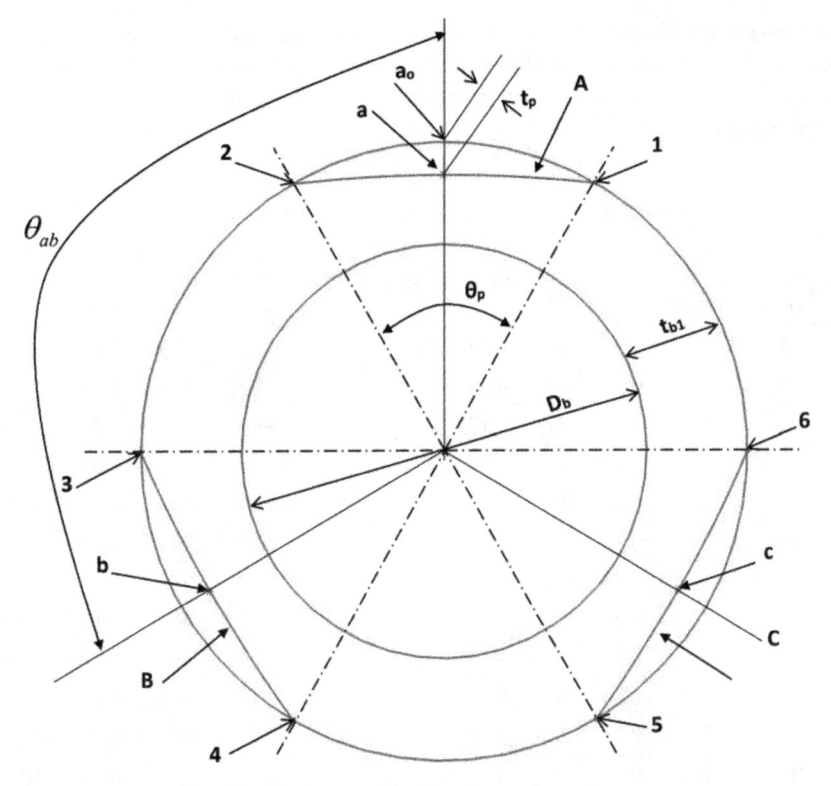

Fig. 17. Shape-morphing bearing pocket geometry

In addition, an important criterion for selecting the optimum pocket design is the deformation of the bearing due to the pressure of the lubricant film during operation. For lower values of the eccentricity ratio (i.e. $\varepsilon = 0.4$) the generated pressures are considerably lower than those produced for bigger ratios (i.e. $\varepsilon = 0.6$ or 0.8). The aim is that this variation does not affect the geometry of the inner surface of the bearing, preventing it from acting as a wave bearing. Based on the sensitivity analysis presented in Sect. 4.4 and the FSI simulations the difference in the deformation of the pockets across the range of the operation eccentricity ratios should not be greater than 10%. So another constraint is $\Delta D \leq 10\%$.

5.2 Pocket Design

The detail design of the shape-morphing bearing was done for $e_w = 0.1$. For different e_w the same procedure can be followed. Different geometrical parameters were tested for the wave bearing of Table 1. More specifically, the deformation of the inner ring was calculated for different values of thicknesses and pocket angles to identify a design that ensures a deviation to the inner ring diameter less than 10%. As the worst case scenario, a pressure of 5.2 MPa (Fig. 16) was applied uniformly to model the pressure of the lubricant film. The results are presented in Table 3. P_{pocket} is the external pressure applies inside the pocket to create a wave-like shape as defined by Eq. (3). D_1 is the maximum deformation when no lubricant pressure is applied (only the external pressure P_{pocket}) and D_2 is the maximum deformation when an additional pressure of 5.2 MPa is applied to the inside surface of the inner ring.

Table 3. Design cases for the shape-morphing wave bearing

No	t_{b1} (mm)	t_p (mm)	t_w	θ_p (°)	θ_w	P_{pocket} (MPa)	D_1 (μm)	D_2 (μm)	ΔD (μm)	$\Delta D\%$
1	4	1	0.25	20	0.17	32	35.5	29.8	0.16	16.1
2	4	1	0.25	60	0.50	8.5	72.3	29.6	0.59	59.0
3	4	2	0.50	20	0.17	24	38.3	29.9	0.22	21.8
4	4	2	0.50	60	0.50	7.3	97.9	30.0	0.69	69.4
5	4	2	0.50	80	0.67	7.5	91.9	29.9	0.67	67.5
6	10	5	0.50	60	0.50	11	52.6	29.2	0.44	44.4
7	15	2	0.13	60	0.50	23	36.9	29.5	0.20	19.9
8	15	4	0.27	40	0.33	41	33.3	29.3	0.12	12.0
9	15	5	0.33	40	0.33	39	33.6	29.4	0.13	12.8
10	15	5	0.33	50	0.42	25	36.6	29.6	0.19	19.1
11	15	5	0.33	60	0.50	19	39.2	29.5	0.25	24.7
12	20	2	0.10	80	0.67	24	36.2	29.6	0.18	18.2
13	20	3	0.15	60	0.50	29.5	34.8	29.5	0.15	15.2
14	20	5	0.25	40	0.33	57	32.3	29.5	0.09	8.8
15	20	8	0.40	40	0.33	52	32.7	29.5	0.10	9.9
16	20	10	0.50	40	0.33	48	33.1	29.5	0.11	10.8
17	25	2	0.08	80	0.67	29	34.4	29.4	0.15	14.7
18	25	2	0.08	100	0.83	28	34.7	29.4	0.15	15.2
19	25	3	0.12	80	0.67	28	34.5	29.2	0.15	15.3
20	25	5	0.20	40	0.33	72	31.8	29.6	0.07	7.1
21	25	5	0.20	50	0.42	46	32.3	29.0	0.10	10.2
22	25	5	0.20	60	0.50	35	34.1	29.8	0.13	12.7
23	25	5	0.20	80	0.67	26.2	35.0	29.2	0.16	16.5
24	25	5	0.20	100	0.83	25.1	35.1	29.1	0.17	17.1
25	25	6	0.24	40	0.33	70	31.5	29.2	0.07	7.4
26	25	8	0.32	40	0.33	68	31.7	29.3	0.08	7.7
27	25	10	0.40	40	0.33	66	32.3	29.7	0.08	8.0
28	25	10	0.40	50	0.42	41	33.6	29.6	0.12	11.8

(*continued*)

Table 3. (*continued*)

No	t_{b1} (mm)	t_p (mm)	t_w	θ_p (°)	θ_w	P_{pocket} (MPa)	D_1 (μm)	D_2 (μm)	ΔD (μm)	$\Delta D\%$
29	25	10	0.40	60	0.50	29.5	34.9	29.4	0.16	15.6
30	25	15	0.60	40	0.33	55	32.7	29.5	0.10	9.9
31	30	2	0.07	100	0.83	32.8	34.1	29.8	0.13	12.6
32	30	4	0.13	60	0.50	42.8	33.2	29.8	0.10	10.3
33	30	5	0.17	40	0.33	83	31.2	29.2	0.06	6.3
34	30	5	0.17	60	0.50	41.2	32.6	29.1	0.11	10.7
35	30	5	0.17	80	0.67	32	34.4	29.9	0.13	13.2
36	30	8	0.27	60	0.50	39	33.3	29.4	0.11	11.5
37	30	10	0.33	40	0.33	80	31.8	29.7	0.07	6.7
38	30	10	0.33	50	0.42	51	32.6	29.5	0.09	9.4
39	30	10	0.33	60	0.50	37	33.3	29.4	0.12	11.7
40	30	15	0.50	40	0.33	72	31.9	29.4	0.08	7.7
41	30	15	0.50	50	0.42	45	32.9	29.3	0.11	11.0
42	30	20	0.67	40	0.33	60	32.6	29.5	0.09	9.4

The results of Table 3 show that as t_w increases (i.e. the thickness of the pocket becomes bigger relative to the inner ring thickness), ΔD increases as well. In addition, θ_w directly affects the wave profile generated inside the bearing and how closely it marches Eq. (3). As it will be shown in Sect. 5.3 a satisfactory wave profile is formed for $\theta_w \approx 0.42 - 0.5$. However, a careful selection of the other geometrical parameters should be made because as θ_w increases, ΔD increases thus limiting even further the permissible values of thicknesses and external pressures. In conclusion, for values of $t_{b1} > 20$ mm, ΔD is within the desired range so most simulations were done for $t_{b1} = 20$–30 mm.

5.3 Detailed Pocket Deformation Analysis

In order to determine the optimum pocket placement to recreate the pressure distribution of wave bearing, a parametric analysis was carried out for 5 separate values of θ_p (40°, 50°, 60°, 80° and 100°) and for different values of t_{b1} and t_p from Table 3. For every case the deformed profile of the inside diameter was compared to the nominal wave profile of the wave bearing. For example, in Fig. 18 the wave profiles for cases 22 and 32 of Table 3 ($\theta_p = 60°$) are plotted against the nominal one. Figure 19 shows the deformation of the inner ring for case 22 as calculated in ANSYS Workbench.

100   C. Kalligeros et al.

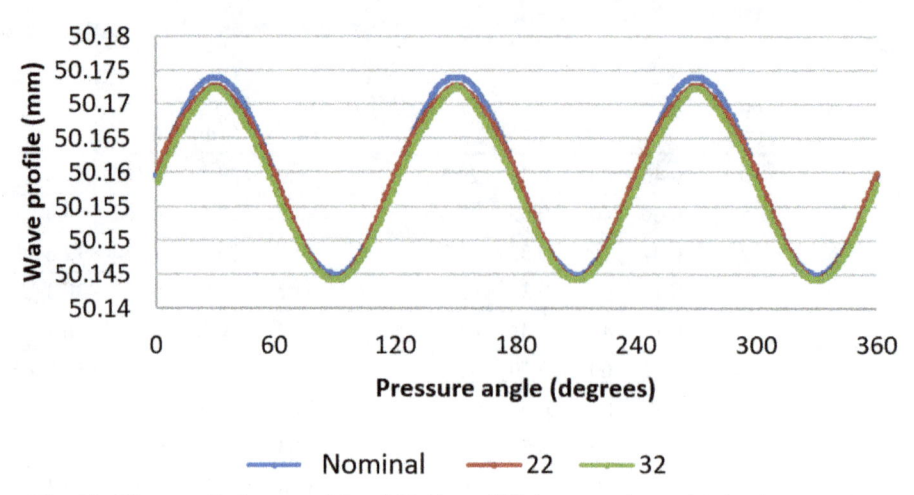

Fig. 18. Wave profile for cases 22 and 32 ($\theta_p = 60°$) in comparison with the theoretical one

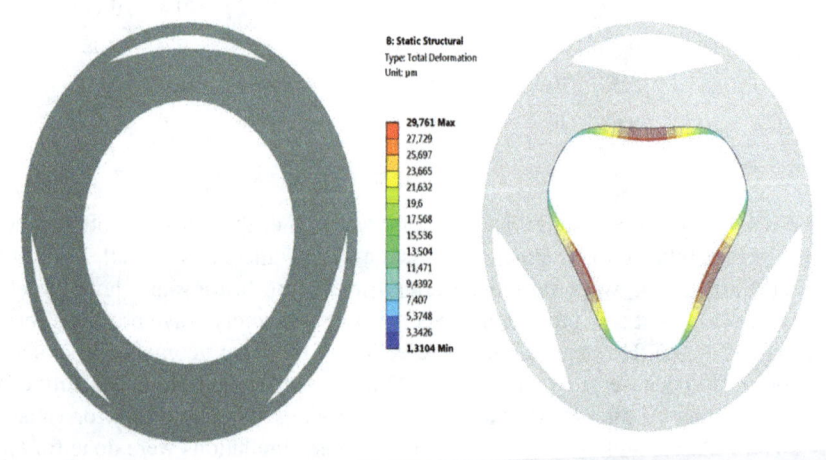

Fig. 19. Deformation of inner ring of shape-morphing wave bearing calculated by ANSYS Workbench

Figure 20 shows the inside profile of the deformed inner ring for four cases (i.e. cases 20, 32, 17 and 31 from Table 3) corresponding to four different values of θ_p (i.e. 40°, 60°, 80° and 100°) respectively.

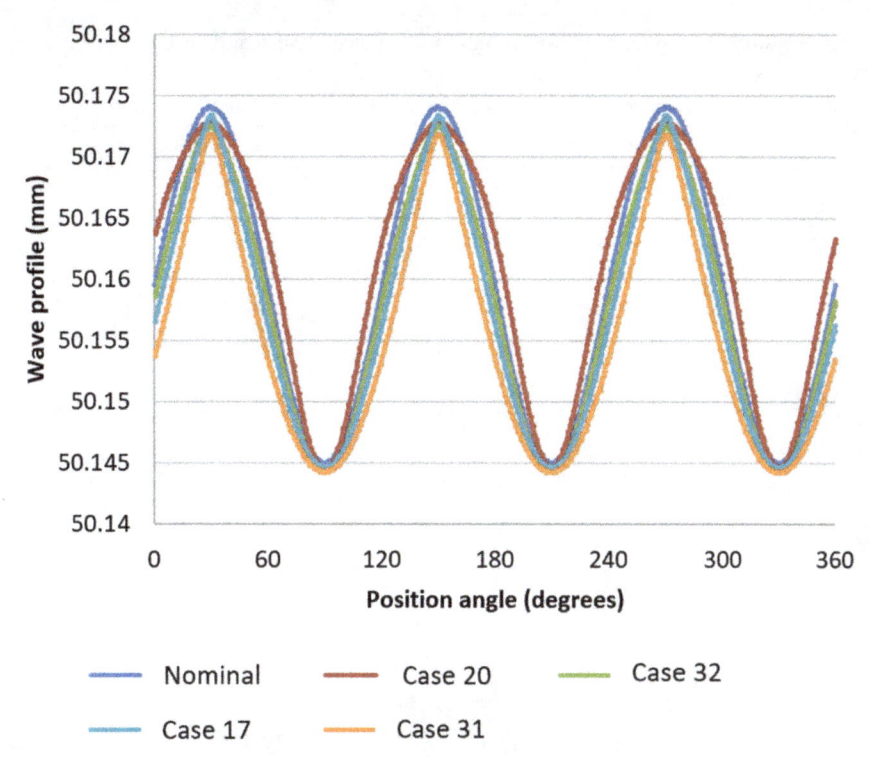

[Chart: Wave profile (mm) vs Position angle (degrees), showing curves for Nominal, Case 20, Case 32, Case 17, Case 31]

Fig. 20. Wave profile comparison between cases 20, 32, 38, 17 and 31 with the nominal one

The minimum points located at 90°, 210° and 330° (i.e. for $D = 50.145$ mm) are very close to the nominal value of θ_p. On the other hand the maxima at 30°, 150° and 270° deviate drpm the nominal value but less than 3 μm. The smaller deviation occurs for $\theta_p = 80°$ when it is approximately 1 μm. The second smaller deviation occurs for $\theta_p = 40°$, while the bigger one for $\theta_p = 100°$. Consequently, the deviation does not increase only as a factor of θ_p but it is dependent on other parameters as well.

By executing a detailed parametric study it was concluded that for value of the ratio $\theta_w \approx 0.42$–0.5 the internal deformed geometry is the closest to the nominal one of the wave bearing. More specifically, case 38 of Table 3 was selected as the best design of the shape-morphing bearing that has the geometrical parameters presented in Table 4. The wave profile achieved for this case is illustrated in Fig. 21.

Table 4. Optimum pocket design for examined wave bearing geometry

Parameter	Value
t_{b1} (mm)	30
t_p (mm)	10
θ_p (deg)	50
ΔD (%)	9.4

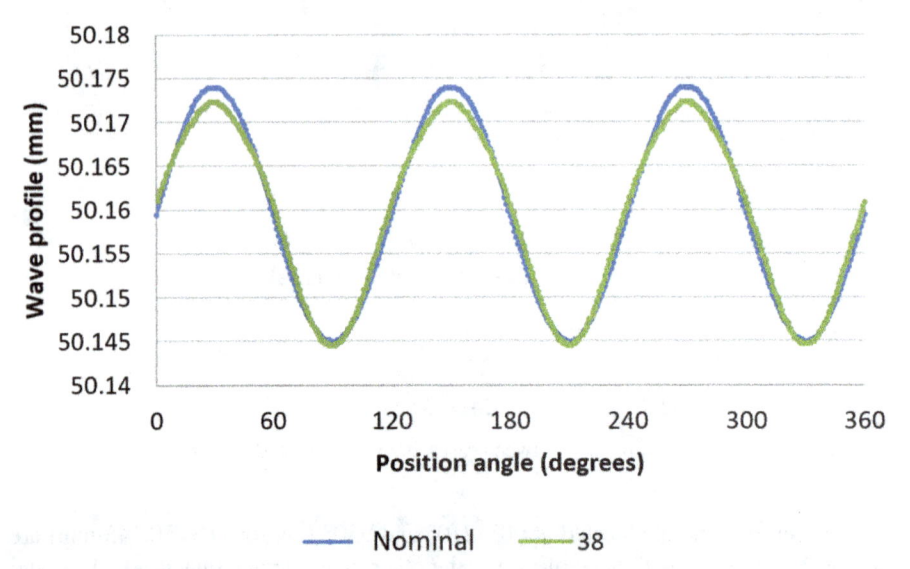

Fig. 21. Wave profile for the optimum shape-morphing design compared with the nominal profile of the wave bearing

5.4 FSI Analysis

The deformation analysis of the shape-morphing bearing presented in Sect. 5.3 assumed a uniform distribution of pressure at the inside surface of the inner ring at the magnitude of the maximum pressure obtained during the CFD analysis. In reality this pressure is developed only on a specific area of the bearing. In Fig. 22 the actual pressure distribution is presented as calculated in ANSYS Fluent after the application of an external positive pressure of 2.5 MPa (as explained in Sect. 4.4). This pressure distribution is imported from Fluent to Static Structural (FSI analysis) to determine the actual deformation profile of the inner ring of the shape-morphing bearing. Figure 23 presents the profile of the inside surface of the bearing of case 38 (Table 3) as calculated from static analysis (uniform pressure assumption) and FSI analysis against the nominal geometry.

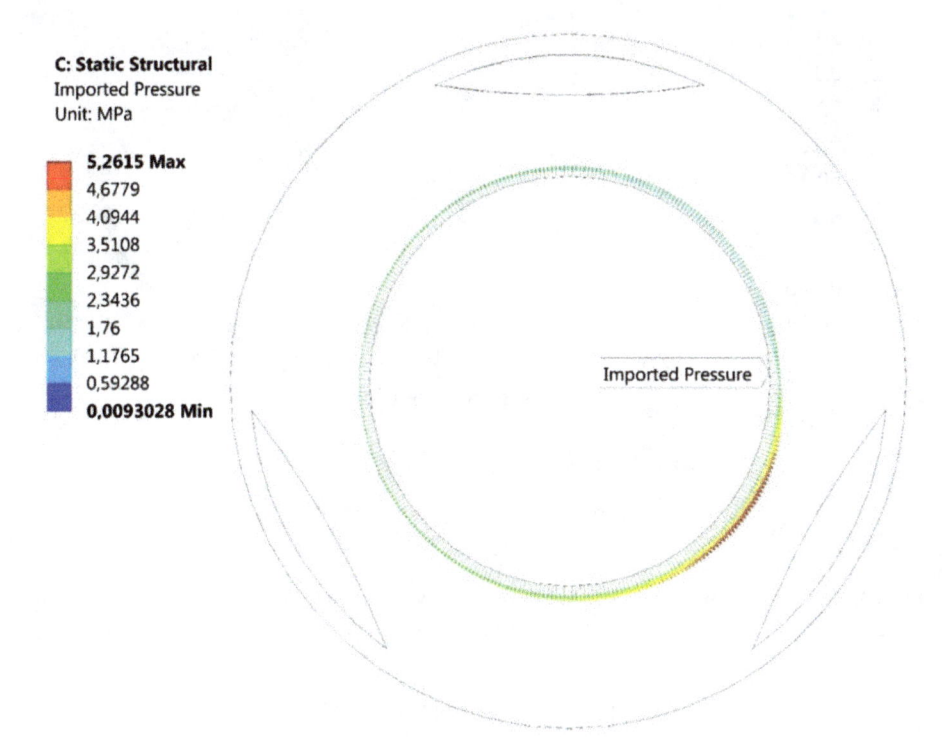

Fig. 22. Pressure distribution of lubricant film after the application of external positive pressure of 2.5 MPa.

The deviation that occurs at the region of maximum pressure has a magnitude of about 1 μm. As shown in Table 5, the major difference compared to the static analysis is that less external pressure (P_{Pocket}) is required to achieve the wave-like deformation and that ΔD decreases from 9.4% to 3.8% because with the exception of the 290°–360° region where high pressures are applied, the pressures are low resulting in less resistance to deformation. Based on the this conclusion more cases of Table 3 where the ΔD is in the range of 10–15% can be used since when examined with the actual pressure distribution the ΔD is expected to decrease below the limit of 10%.

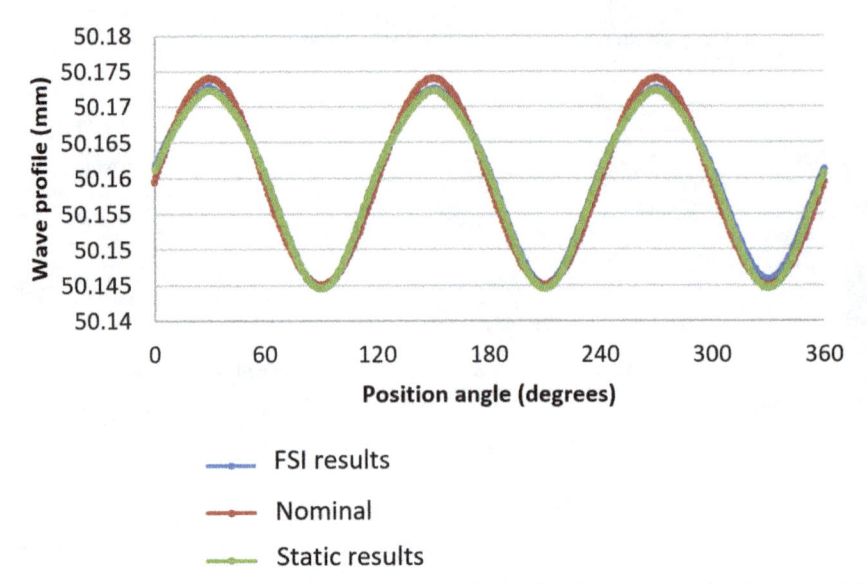

Fig. 23. Comparison of wave profile for the nominal wave bearing (orange), the shape-morphing bearing after static analysis (grey) and the shape-morphing bearing after FSI analysis (blue)

Table 5. Pocket design based on Static and FSI analysis

	t_{b1} (mm)	t_p (mm)	t_w	θ_p (°)	θ_w	P_{pocket} (MPa)	D_1 (μm)	D_2 (μm)	ΔD (μm)	$\Delta D\%$
Static Analysis	30	10	0.33	50	0.42	51	32.6	29.5	0.09	9.4
FSI analysis	30	10	0,33	50	0.42	48	30.7	29.5	0.04	3.8

If a bigger external pressure is applied in the pocket where the bearing withstands the maximum lubricant pressures, the deviation of approximately 1 μm shown in Fig. 23 becomes zero. More specifically, if approximately 1.5 MPa additional pressure is applied to the third fluid chamber (Fig. 24), the inner surface of the bearing of the real model is even closer to the inner surface of the nominal wave bearing as shown in Fig. 25.

At points where $R = 50.174$ mm the deviation of about 3 μm still remains. Although this deviation is not negligible it is difficult to be reduced with the current structure of the pockets. The adoption of more pockets may help overcoming this problem, but this is beyond the scope of this work. But since in Sect. 4.4 it was shown that small deviations like this do not have a significant effect on pressure distribution, it is not expected to prevent the shape-morphing bearing to act as a wave bearing.

C: Static Structural
Static Structural

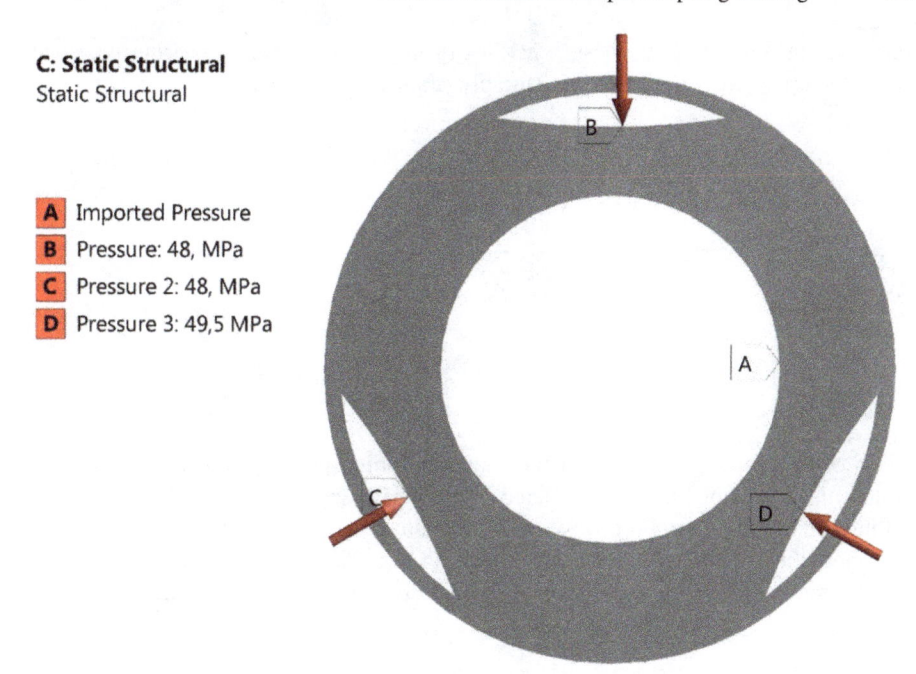

A Imported Pressure
B Pressure: 48, MPa
C Pressure 2: 48, MPa
D Pressure 3: 49,5 MPa

Fig. 24. External pressures applied to pockets

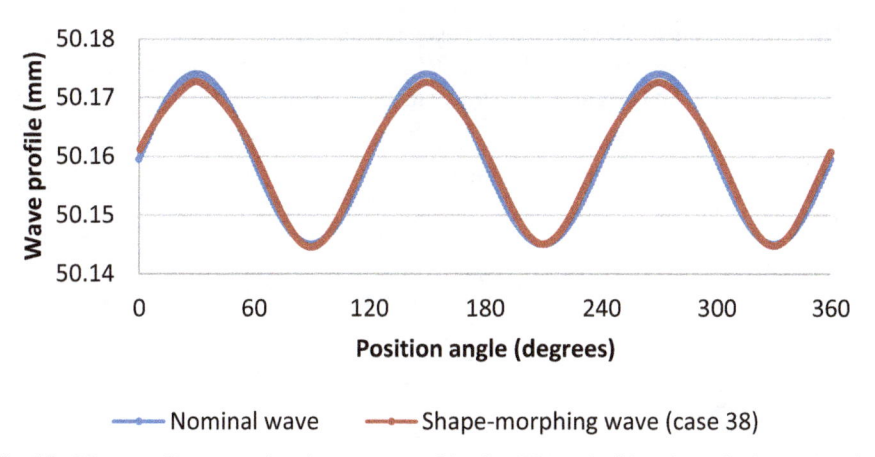

Fig. 25. Wave profile comparison between case 38 (after FSI analysis) and nominal wave bearing geometry

6 Dynamic Analysis

The role of the shape-morphing wave bearing in a rotating system is twofold. On the one hand, they can adjust the levels of shaft oscillations by varying their stiffness characteristics. On the other hand, they can drive the system into anti-resonance, in which case, due to the presence of damping effects, the amplitude of oscillations becomes very small. In this section, both these functions of the shape-morphing bearing waves are presented on an arrangement that is suitably adapted for each case.

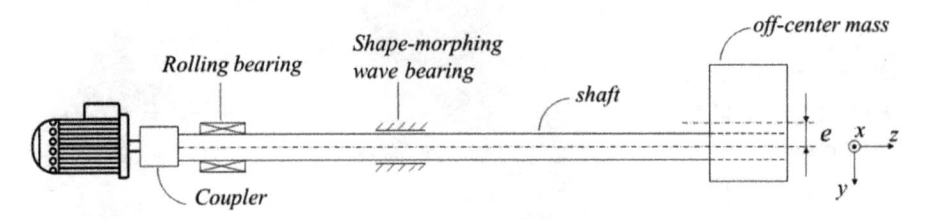

Fig. 26. Dynamic simulation layout

In order to study the effect of the shape-morphing bearing to a rotating dynamic system the layout of Fig. 26 is selected. A motor rotates a shaft that is mounted on a rolling and a shape-morphing wave bearing and has an off-center mass on the one end. The modelling of the arrangement is shown in Fig. 27. The shaft with the off-center mass is modelled as a mass-spring system with two degrees of freedom. The bending stiffness of the system is given by:

$$K_s = \begin{bmatrix} k_s & 0 \\ 0 & k_s \end{bmatrix}, k_s = \frac{3EI}{L^3} \quad (5)$$

The torsional stiffness of the shaft is neglected since there is no load in the assembly that prevents the free rotation of the shaft. The damping of the sub-assembly itself is also negligible since it is significantly lower than that of the wave bearing.

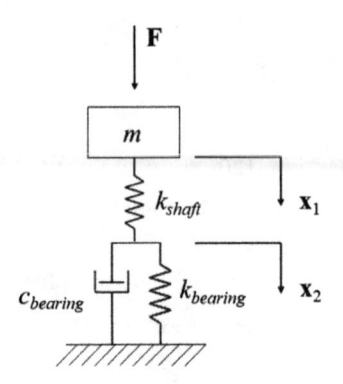

Fig. 27. Modeling of dynamic simulation layout

The shape-morphing wave bearing is modelled as a spring-damper system. The stiffness characteristics of the bearing are calculated by the FEA presented in Sects. 4 and 5, while its damping coefficients were calculated implementing the analysis presented in [17].

As mentioned earlier, the role of the shape-morphing wave bearing can be twofold. The first one relates to its ability to regulate the level of oscillations of a system by varying the pressure applied externally to the pockets of bearing and thus the geometry of the inner ring. Modifying the geometry of the bearing causes a change in the wave amplitude and the pressure distribution to the lubricant film. In this way, a change of the bearing's stiffness and damping occurs affecting the dynamic response of the whole system.

Figure 28 shows the maximum amplitude of the oscillations arising on the shaft of the arrangement shown in Fig. 26, with the geometrical and dynamic characteristics given in Table 6. As can be seen from Fig. 28, increasing the external pressure to the shape-morphing bearing leads to a reduction of shaft oscillations. In this particular case, a reduction of about 100% (exact reduction of 94%) can be achieved.

Table 6. Layout parameters for oscillation reduction

Parameter	Value
Rotation speed (rpm)	1500
Shaft diameter (mm)	20
Projection length (mm)	200
Cam diameter (mm)	30
Cam length (mm)	20
Eccentricity e (mm)	1
Modulus of elasticity (GPa)	210
Density (kg/m^3)	7800
Viscosity of lubricant (Pa s)	0.0166

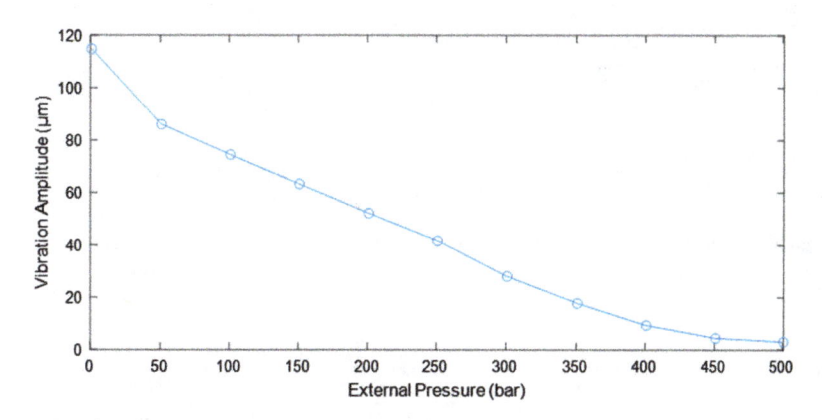

Fig. 28. Maximum shaft oscillation amplitude as a function of pocket pressure

The second function of the shape-morphing wave bearing is to lead the system to anti-resonance response. With an appropriate change in the stiffness of the bearings, the system is excited, but the amplitude of the oscillation remains low due to the damping action of the bearings. In this way NVH emissions remain low.

For the geometrical parameters of Table 7, the eigenfrequency values of the system for different values of external pressure are shown in Fig. 29.

Table 7. Layout parameters for anti-resonance response

Parameter	Value
Rotation speed (rpm)	1500
Shaft diameter (mm)	20
Projection length (mm)	1800
Cam diameter (mm)	50
Cam length (mm)	200
Eccentricity e (mm)	0.015
Modulus of elasticity (GPa)	210
Density (kg/m^3)	7800
Viscosity of lubricant (Pa s)	0.0166

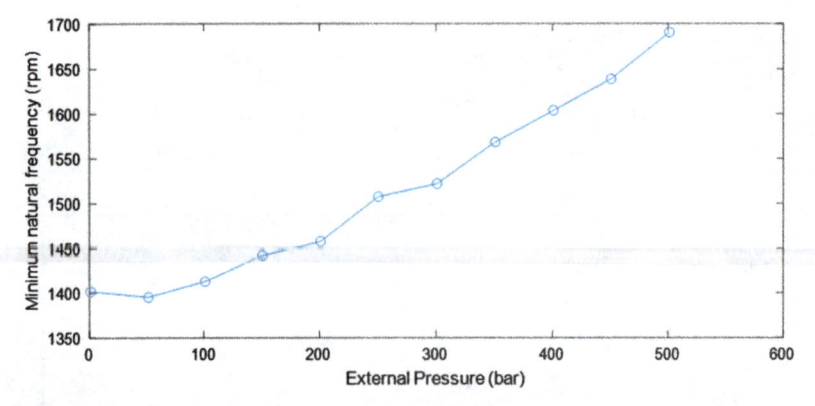

Fig. 29. Minimum eignefrequency of the rotating system as a function of pocket pressure

7 Conclusions

In this work a novel shape-morphing journal bearing was presented that can actively modify its inside surface to transform into a three-wave journal bearing. The shape-morphing bearing consists of two cylindrical rings one inside the other. The outside diameter of the inner ring is not exactly cylindrical creating three pockets when the two rings are assembled. Inside this pockets external hydraulic pressure is applied so the inside surface of the inner ring is deformed to a wave-like shape that can operate as a wave bearing. By modifying the external pressures applied into the pockets, the wave amplitude change along with the dynamic characteristics of the bearing, offering the ability to control the dynamic response of the system.

In order to design the pockets of the shape-morphing wave bearing CFD and FSI analyses were implemented. It was shown that it is possible to deform the inside surface of the inner ring to a profile that closely resembles the nominal profile of the wave bearing. A sensitivity analysis showed the occurring deviations do not have a significant effect on the pressure distribution developed on the lubricant film and consequently the shape-morphing bearing can operate reliably as a wave bearing.

A dynamic analysis was carried out and revealed the capabilities of the shape morphing bearing to control the dynamic response of a rotating system. The two main functions examined were the adjustment of the level of shaft oscillations and the capability of achieving anti-resonance effects. The analysis revealed that for the examined case a reduction of shaft oscillations by 100% can be achieved. In addition, anti-resonance can occur leading to dampening of oscillations especially for systems with low-stiffness characteristics.

References

1. Dimofte, F.: A waved journal bearing concept-evaluating steady-state and dynamic performance with a potential active control alternative. Presented at the ASME 1993 Design Technical Conferences 18 March 2021 (2021). https://doi.org/10.1115/DETC1993-0186
2. Dimofte, F.: Wave journal bearing with compressible lubricant—part I: the wave bearing concept and a comparison to the plain circular bearing. Tribol. Trans. **38**, 153–160 (1995). https://doi.org/10.1080/10402009508983391
3. Dimofte, F.: Wave journal bearing with compressible lubricant—part II: a comparison of the wave bearing with a wave-groove bearing and a lobe bearing. Tribol. Trans. **38**, 364–372 (1995). https://doi.org/10.1080/10402009508983416
4. Dimofte, F, Addy Jr., H. E.: Advanced earth-to-orbit propulsion technology. In: 1994 Proceedings of a Conference Held at NASA George C. Marshall Space Flight Center, Marshall Space Flight Center, 17–19 May 1994. National Aeronautics and Space Administration, Marshall Space Flight Center (1994)
5. Dimofte, F., Proctor, M.P., Fleming, D.P., Keith, T.G.: Wave fluid film bearing tests for an aviation gearbox (2000)
6. Guo, Y., Eritenel, T., Ericson, T.M., Parker, R.G.: Vibro-acoustic propagation of gear dynamics in a gear-bearing-housing system. J. Sound Vib. **333**, 5762–5785 (2014). https://doi.org/10.1016/j.jsv.2014.05.055
7. Yang, B., et al.: An investigation on the static performance of hydrodynamic wave journal bearings. In: 2016 IEEE International Conference on Mechatronics and Automation, pp. 61–67 (2016). https://doi.org/10.1109/ICMA.2016.7558535

8. Yang, B., et al.: Parameter study on dynamic characteristics of wave journal bearings. In: 2018 IEEE International Conference on Mechatronics and Automation (ICMA), pp. 220–226 (2018). https://doi.org/10.1109/ICMA.2018.8484278
9. Ene, N.M., Dimofte, F., Keith, T.G.: A stability analysis for a hydrodynamic three-wave journal bearing. Tribol. Int. **41**, 434–442 (2008). https://doi.org/10.1016/j.triboint.2007.10.002
10. Ene, N.M., Dimofte, F., Keith, T.G., Handschuh, R.F.: The robust stability of a wave bearing. Presented at the World Tribology Congress III, 17 November 2008 (2008). https://doi.org/10.1115/WTC2005-64084
11. Ene, N.M., Dimofte, F.: Effect of fluid film wave bearings on attenuation of gear mesh noise and vibration. Tribol. Int. **53**, 108–114 (2012). https://doi.org/10.1016/j.triboint.2012.04.016
12. Chasalevris, A., Guignier, G.: Alignment and rotordynamic optimization of turbine shaft trains using adjustable bearings in real-time operation. Proc. Inst. Mech. Eng. C J. Mech. Eng. Sci. **233**, 2379–2399 (2019). https://doi.org/10.1177/0954406218791636
13. Chasalevris, A., Dohnal, F.: Improving stability and operation of turbine rotors using adjustable journal bearings. Tribol. Int. **104**, 369–382 (2016). https://doi.org/10.1016/j.triboint.2016.06.022
14. Ene, N.M.: Stability and thermohydrodynamic investigations of wave journal bearings. The University of Toledo (2008)
15. Dhande, D., Pande, D.W., Chatarkar, V.: Analysis of hydrodynamic journal bearing using fluid structure interaction approach. Int. J. Eng. Trends Technol. **4**(8) (2008)
16. Bompos, D.A., Nikolakopoulos, P.G.: CFD simulation of magnetorheological fluid journal bearings. Simul. Modell. Pract. Theory **19**, 1035–1060 (2011). https://doi.org/10.1016/j.simpat.2011.01.001
17. Chasalevris, A.: Analytical evaluation of the static and dynamic characteristics of three-lobe journal bearings with finite length. J. Tribol. **137** (2015). https://doi.org/10.1115/1.4030023

Application of Squeeze Film Dampers

Edoardo Gheller[✉][iD], Steven Chatterton[iD], Andrea Vania[iD], and Paolo Pennacchi[iD]

Department of Mechanical Engineering, Politecnico di Milano, Via G. la Masa 1, 20156 Milan, Italy
{edoardo.gheller,steven.chatterton,andrea.vania, paolo.pennacchi}@polimi.it

Abstract. The level of the vibrations and the presence of instability are the two most critical aspects regarding the operations of turbomachinery. To cope with this issues that may compromise the operation of the machines, squeeze film dampers (SFD) are often used in many industrial applications. Unfortunately, many complex phenomena characterize the dynamic behavior of these components and determine the high complexity of the modeling of these components. The most relevant phenomena involved in the characterization of SFDs are individuated after a comprehensive investigation of the state of the art. Among them, the oil film cavitation, the air ingestion, and the effect of the inertia are introduced. A modeling strategy based on the Reynolds equation is then presented. The boundary conditions to be adopted for the feeding and discharging of oil are investigated and implemented. Eventually, the finite difference model is applied to a practical example to evaluate the possibility to minimize the vibration level and to reduce the effect of the instability if a SFD is added to a rotodynamic system. Meaningful information about the modeling of SFDs is provided in this work. The critical aspects of these components and their modeling are highlighted and discussed.

Keywords: Squeeze Film Damper · Instability · Vibrations

1 Introduction

In mechanical engineering high levels of vibrations are problematic in all fields, especially in rotordynamics. To increase the productivity and the efficiency of rotating machines, the load that applied to these mechanical components are becoming more and more remarkable. Also considering the increase of the operation speeds, to guarantee safe long operation times is fundamental to reduce the vibrations. In rotordynamics, the typical problems to be dealt with are the high levels of steady state synchronous vibration and the subsynchronous rotor instabilities. The first one is usually dependent on excessive unbalance or if the machine is operated close to a critical speed. On the contrary, the second one is linked to the presence of some instability sources, connected to cross-coupling effects arising, for example, in bearings or seals. Moreover, the level of the vibration when a critical speed is crossed during a runup, or a rundown may also be detrimental to the operation of the machine. Given these considerations, the dynamic response of the system should benefit from the addition of damping. When the system is supported by rolling element bearings, the damping introduced in the system may be insufficient. Therefore, squeeze film dampers (SFDs can be considered to this aim. Both the dissipation of the vibrational energy and the improvement of the dynamic stability of the rotor-bearing system can be improved by these components.

The most widespread design for such components is the one coupled with rolling element bearings, as shown in Fig. 1.

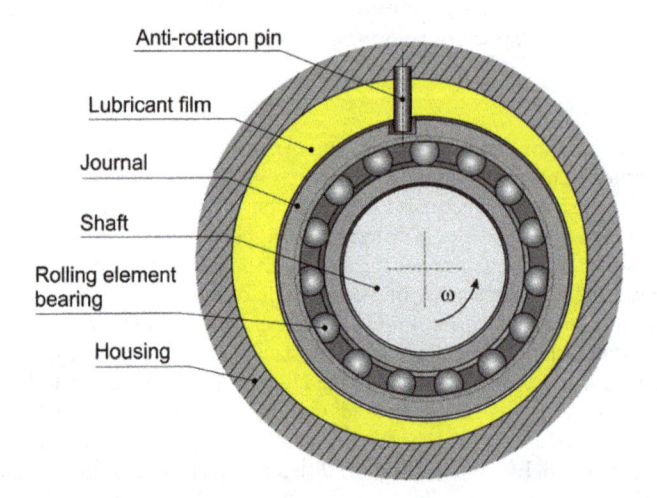

Fig. 1. SFD schematization.

A rolling element bearing supports the shaft. The coupling of the two elements is often referred as journal. The external ring of the bearing moves together with the shaft and the lubricant film, placed between the external surface of the journal and the housing, is "squeezed". The damping effect is generated by the large dynamic forces generated by the large dynamic pressures generated by the squeeze. An anti-rotation pin may be adopted to avoid the transmission of the spinning motion from the shaft to the oil. Only lateral displacement is possible. In other words, the journal cannot spin around its axis of symmetry but only translate. The bearing presence determines the decoupling of the journal motion from the shaft spinning. The shaft kinematics are explained more in the detail in Sect. 2.1.

Most of the times, supports are applied to sustain the journal at the runup until the vibration amplitude is high enough to guarantee the detachment between the casing and the journal. Therefore, a supporting structure, such as O-rings and squirrel cages, is applied to reduce the risk of impacts. Then they avoid the presence of strong non-linearity determined by the detachment of the journal from the casing. The stiffness of the supporting structure is indeed one fundamental parameter for the optimal operation of the SFD. If the stiffness of the supporting structure is over dimensioned, the relative motion between the journal and the cage will be limited reducing the squeezing of the lubricant film. On the contrary, if the supporting structure is not stiff enough, the SFD can behave as a non-supported one and may be subject to impacts and damages, [1, 2].

SFDs are designed to introduce in the system the optimal level of damping that is strongly dependent on the application. For example, SFDs characterized by lower damping may reduce more the amplitude of the than the ones characterized by higher damping.

The ideal level of damping needed by a machine generally depends on the type of the excitations, the operating conditions, and the dynamic characteristics of the system [3, 4]. In the literature, the modeling of the dynamic behavior of SFDs is performed with numerous models characterized by distinct levels of complexity. The 1D Reynolds equation was the first one considered for SFDs with length to diameter ratios lower than 0.25 and if no sealing mechanism is adopted, [5]. Linearized stiffness and damping coefficients are adopted. However, if the spinning motion is not considered, no stiffness is introduced in the system by the SFD. When the hypothesis of infinitely long bearing is valid, another form of the 1D Reynolds equation is adopted [5]. For both approximations, the resulting equations can be analytically solved, and numerous examples of force coefficients may be found, [2, 3].

The shaft motion needs to be modeled to characterize the dynamic behavior of SFDs.

Two different approaches can be followed: circular synchronous orbits, centered or with a statically eccentric, or small amplitude motions about a static displaced center. The first one is usually considered to investigate the response to unbalance, while the second one is considered to analyze the stability of the system, [6].

The SFD clearance and length to diameter ration results as fundamental parameters to characterize the operation of the SFD depending on the level of the vibration, [1–3]. The simplicity is the main advantage of the models based on the 1-D Reynolds equation but, unfortunately, the resulting predictions are accurate only for a reduced range of operating conditions and for simplified geometrical configurations.

Many complex phenomena, not captured by the previously mentioned models, affect the dynamic response of SFDs such as the inertia, the cavitation, the air entrainment, and complex geometrical features.

In the derivation of the Reynolds equation the presence of inertia is usually neglected. However, for large clearances and amplitudes of motions and frequencies, the dynamic pressurization of the lubricant film generates an added mass that was experimentally found in [7] to be of the same order of magnitude of the mass of the whole SFD. Many authors deal with the effect of inertia. In [8], assume that for moderate values of the Squeeze Reynolds Number ($Re = \frac{\rho \omega c l^2}{\mu} \leq 10$ with ρ being the density and μ being viscosity of the fluid, cl the SFD clearance and ω the vibration frequency) the fluid inertia does not affect the shape of the purely viscous velocity profile. Only the effect of temporal inertia is considered by the authors while both the effects of temporal and convective inertia are considered in [9].

Moreover, the presence of cavitation is considered as the main reason why predictions made with the simplified modes used in [1, 3] do not agree well with the experimental results. Zeidan and Vance in [10] report five different cavitation regimes: absence of cavitation, cavitation bubble following the journal, mixture of oil and air, vapor cavitation, combination of vapor and gaseous cavitation. The second cavitation regime may be assumed as a transient condition, present only at small whirling frequencies that eventually turns into the third one at higher speeds. The most common regimes that are encountered are the third and the fourth that sometimes appear together as combined. In [11], Diaz and San Andrés focused on air ingestion and vapor cavitation. A SFD was tested by the authors in open-ends and in fully flooded configuration, considering different whirling frequencies and pressures of supply oil. The dynamic pression generated was measured in time and the differences between the two cavitation regimes was shown. When vapor cavitation occurs, the measured pressure evolution did not change for the different revolutions while, for air entrainment, the pressure profiles changed significantly from one revolution to the other. The differences showed by the two cavitation regimes often pushed the authors to treat and model them separately. The π-film model, also known as Gumbel condition, can be considered to model the vapor cavitation. In this case, the pressure is considered zero in the region where it assumes negative values. Half of the circumferential length of the SFD is therefore characterized by a ruptured lubricant film, see Fig. 2. More refined cavitation algorithms are the *Elrod's cavitation algorithm*, [12] and the *Linear Complementarity Problem* (LCP).

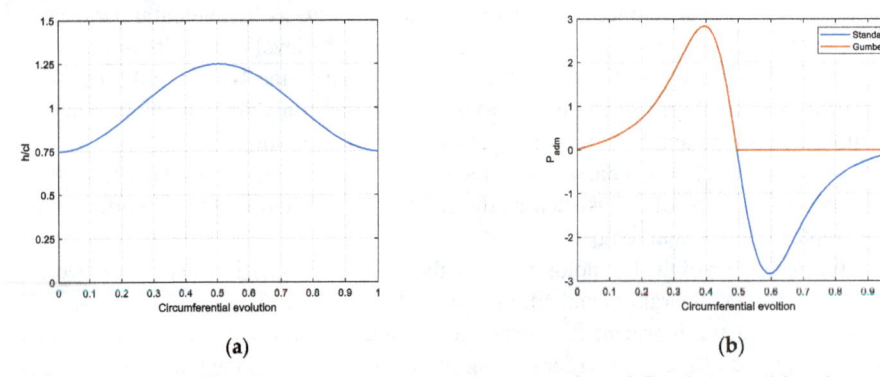

Fig. 2. Evolution of oil film thickness (**a**) and representation of Gumbel condition for vapor cavitation (**b**).

For what regards air ingestion, in [11, 13], and [14] its effect is experimentally investigated. When this phenomenon occurs, air is "sucked" inside the SFD, and, after some time, a fine mixture of oil and air bubbles is formed. Air bubbles are present also in the high-pressure zone and the variability of the values of the pressure peaks shown in [11] can be explained by their presence. Diaz [15] presented a detailed strategy, supported by numerous experimental results, to model the effect of the air entrainment in the 2-D

Reynolds equation. The oil-air bubbly mixture is considered to be homogeneous and the distribution of the air content inside the mixture is calculated starting from a reference value. In the experimental campaign conducted in [13], and [14], the SFD is directly fed by a controlled mixture of oil and air. On the contrary, in industrial applications, pure oil is fed to the SFD, and air enters from the region where the lubricant is discharged. So, it is necessary to estimate the value of reference air ingested. A model to predict the air entrainment for short open-ends SFDs is presented in [16]. Mendez then adapted the model of Diaz et al. [17] to finite length SFDs. Both models presented in [16] and [17] are based on the static form of the Rayleigh-Plesset equation to model the effect of air bubbles in the oil for open-ends SFDs. A more complex modeling strategy, where the complete form of the Rayleigh-Plesset equation is adopted, is presented by Gehannin et al. in [18]. Moreover, the differences between the modeling with the complete and the simplified form of the Rayleigh-Plesset equation are shown by the same authors.

A thorough experimental and numerical investigation on the effect of different geometrical configurations on the dynamic characteristics of SFDs is shown in [19]. The effect on the force coefficients of the SFD clearance, length, hole feeding and groove feeding, number and disposition of feeding holes, open ends and sealing ends, whirl orbit amplitude, and shape of orbit is investigated.

A model based on the 2D Reynolds equation is briefly introduced and validated with experimental and numerical results present in the literature. Then, an unbalanced centrifugal compressor is considered to evaluate the reduction of the level of the vibration with the addition of a SFD to the system. The effectiveness of the application of the SFD in correcting the instability of the machine is also investigated. The authors are aware that more accurate models based on the bulk-flow equations [20], and computational fluid dynamics [21–24] are described in the literature. Both modeling strategies are characterized by higher precision of the results. Unfortunately, both the modeling and computational efforts are higher.

2 SFD Model

The 2D Reynolds equation, discretized with the finite difference approach, is adopted in the model proposed. To the standard equation, two extra terms are added to model the ai ingestion and inertia.

2.1 Oil Film Modeling

The dynamic behavior of SFDs is studied considering circular orbit motions of the journal, whether centered (see Fig. 3) or not, or small perturbations around the equilibrium position. The model proposed is developed for circular orbits. However, the model can be easily adapted to deal with non-circular orbits and small motions around the equilibrium position when it is possible to describe the evolution of the lubricant film thickness in time.

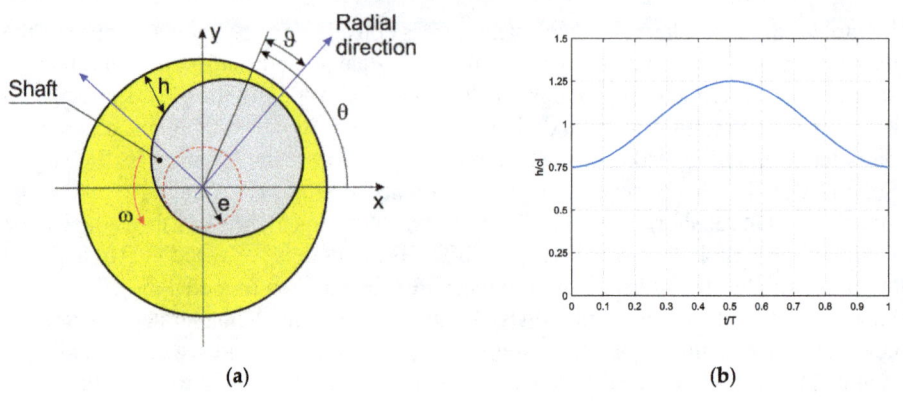

Fig. 3. Representation of circular centered orbit (**a**) and relative evolution of oil film thickness (**b**)

The rotation ϑ in the relative frame of reference and the rotation θ in the absolute frame of reference (x − y) are related as follows:

$$\theta = \vartheta + \omega t \tag{1}$$

If the fixed reference system is considered, the variation in time and space domains of the oil film thickness can be written as:

$$h(\theta, t) = cl - (e \cos \omega t + e_s \cos \theta_s)\cos \theta - (e \sin \omega t + e_s \sin \theta_s)\sin \theta \tag{2}$$

where e_s and θ_s are the amplitude and phase of the static eccentricity. e is the orbit radius.

At each time instant:

$$\frac{\partial h}{\partial t} = -\omega \frac{\partial h}{\partial \vartheta} = -\omega \frac{\partial h}{\partial \theta} \tag{3}$$

Equation (3) allows to simplify the time derivatives in spatial derivatives if the orbiting frequency remains constant in time and if the feeding and sealing system are not considered. Therefore, the transformation proposed in Eq. (3) allows to reduce the calculation time since the pressure distribution at one orbit location is representative of the SFD behavior for the whole orbit.

2.2 Reynolds Equation

The equations to describe the dynamic behavior of a viscous Newtonian fluid are the 3-D Navier-Stokes equations:

$$\frac{\partial \rho}{\partial t} + \nabla \cdot \left(\rho \vec{V} \right) = 0 \tag{4}$$

$$\rho \left(\frac{\partial \vec{V}}{\partial t} + \vec{V} \cdot \nabla \left(\vec{V} \right) \right) = -\nabla P + \nabla \cdot \left(\mu \nabla \vec{V} \right) + \nabla \left(-\frac{2\mu}{3} \nabla \cdot \vec{V} \right) + \rho g \tag{5}$$

where (4) is the continuity equation and (5) are the conservation of momentum equations within the flow boundary.

For the SFD application, some simplifications can be considered. For example, fluid density ρ is considered constant, fluid kinematic viscosity is constant, inertia and body forces are neglected, fluid flow is considered laminar.

Since the fluid film thickness is small and the geometry of the SFD, the curvature of the surfaces can be neglected, and the surfaces can be considered planar. Moreover, the oil fil film thickness is about three orders of magnitudes lower than the axial and circumferential dimensions of the SFD sot the velocity gradients along the latter two dimensions are negligible. Finally:

$$\frac{\partial}{\partial x}\left(h^3 \frac{\partial P}{\partial x}\right) + \frac{\partial}{\partial z}\left(h^3 \frac{\partial P}{\partial z}\right) = 12\mu \frac{\partial}{\partial t}(h) \qquad (6)$$

If the whirling frequency is constant, Eq. (3) can be substituted inside Eq. (6). On the other hand, if the effect of inertia is not neglected, the equations of momentum are different. In [25], the authors state that it is legitimate to hypothesize that the shape of the purely viscous velocity profiles is not affected by the fluid inertia, at least for moderate values of Reynolds number (Re). Moreover, considering average quantities in the flow equations the wall shear stress differences are approximated, [26].

In this work, a similar approach as in [26] is adopted. A single Reynolds-like equation is adopted in which the temporal inertia effect is added. Convective inertia effect is considered negligible as in [25]. In cylindrical coordinates:

$$\frac{\partial}{R\partial \theta}\left(\frac{h^3}{12\mu} \frac{\partial P}{R\partial \theta}\right) + \frac{\partial}{\partial y}\left(\frac{h^3}{12\mu} \frac{\partial P}{\partial y}\right) = \frac{\partial}{\partial t}(h) + \frac{Reh^2}{12\omega c l^2} \frac{\partial^2 h}{\partial t^2} \qquad (7)$$

The comparison between the pressures, normalized w.r.t the reference ambient pressure, obtained with and without the inertial term is shown in Fig. 4. The pressure obtained considering the inertia term remains larger than the ambient reference for a longer time and is also flatter than the pressure calculated with the standard Reynolds equation. For the case with inertia, both the maximum and minimum values are reduced and slightly shifted. In this case the Reynolds number is about 3.5. From the classical Reynolds equation, the force obtained from the pressure distribution is purely tangential and opposes the vibration of the journal. On the other hand, if the temporal inertia term is considered, the resulting force is characterized both by a radial and tangential component. The first one counterbalance the radial acceleration introduced by the inertia term.

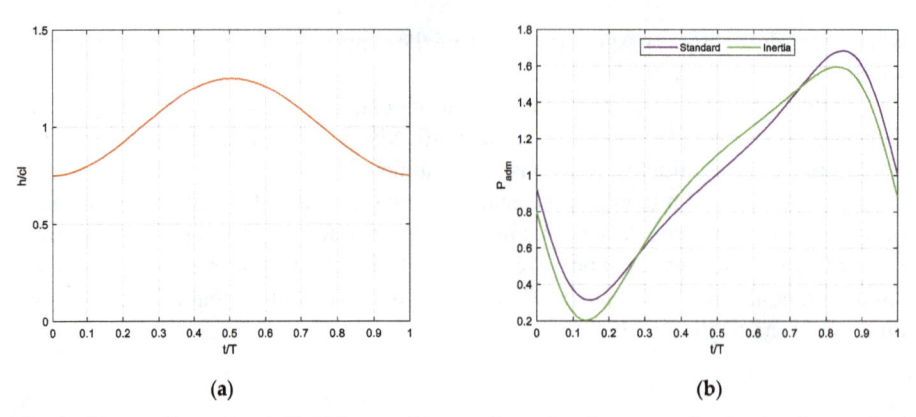

Fig. 4. (a) non-dimensional oil thickness; (b) non-dimensional pressure distribution for standard and Reynolds equation with inertia.

2.3 Air Ingestion

Air entrainment is modeled considering the same approach introduced by the authors in [16]. The presence of the air bubbles in the lubricant affects the density and viscosity of the lubricant. The Reynolds equation becomes:

$$\frac{\partial}{R\partial\theta}\left(\frac{\rho h^3}{12\mu}\frac{\partial P}{R\partial\theta}\right) + \frac{\partial}{\partial z}\left(\frac{\rho h^3}{12\mu}\frac{\partial P}{\partial z}\right) = \frac{\partial}{\partial t}(\rho h) + \frac{Re h^2}{12\omega c^2}\frac{\partial^2 \rho h}{\partial t^2} \quad (8)$$

$$\rho = (1-\beta)\rho_L \quad (9)$$

$$\mu = (1-\beta)\mu_L \quad (10)$$

$$\beta = \frac{1}{1 + \frac{P(x,t)-P_v}{P_{G\sigma}}(\frac{1}{\beta_0} - 1)} \quad (11)$$

β is the air-mixture volume fraction, β_0 is the reference value for β, P_v is the vapor cavitation pressure, $P_{G\sigma}$ is the pressure of the air bubble for the critical radius, and μ_L and ρ_L are the viscosity and density for the pure oil.

The characterization of β_0 is necessary to evaluate the air entrainment level. A simplified model, based on the balance of the flow rates for short SFDs is presented in [16]. Unfortunately, the short bearing hypothesis is applicable only for $L/D < 0.2$. An approach valid for finite length SFD is introduced in [17]. The authors propose to estimate numerically the volumetric flowrate of air entering at the sides of the SFD to evaluate β_0. This approach is then repeated iteratively to update the dynamic pressure distribution until the convergence of the reference air-volume fraction. In this paper, the approach proposed in [17] is adopted.

2.4 Negative Pressure Zone

As previously mentioned, vapor cavitation can be modeled according to different algorithms. In this work, different cavitation models have been tested. Two are based on the

LCP, [27] and [28], and one is based on the Elrod's cavitation algorithm, [29]. All the models tested are effective in calculating the pressure distribution when the standard Reynolds equation is considered. On the contrary, some difficulties were encountered when the inertia contribution is considered. The same holds for the modeling of the air ingestion.

Therefore, a simplified approach is adopted. The dynamic pressure is imposed to be equal to the vapor cavitation value when it assumes lower values than that threshold during the iterative solution of the Reynolds equation. As of now, this simplification allows the authors to include in the analysis some more interesting phenomena like the inertia effect and the air entrainment.

2.5 Geometry

The cylindrical geometry of the damper il opened in a bidimensional plane, see Fig. 5.

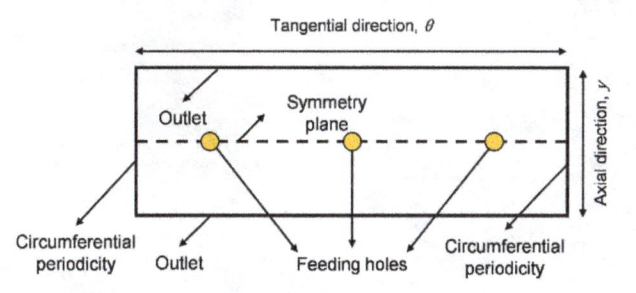

Fig. 5. 2D representation of the SFD.

The mesh considered for the spatial discretization is structured. The approach followed for the boundary conditions is presented in the sub-sections below.

2.6 Boundary Conditions

Inlet
The inlet boundary condition can be modeled in many ways. As a simplified approach, half of the SFD can be considered by applying the symmetry boundary condition:

$$\left.\frac{\partial P}{\partial y}\right|_{symmetry} = 0 \qquad (12)$$

If Eq. (12) is adopted, the feeding system is not modeled.

On the contrary, if the feeding system is considered, the inlet flowrate is imposed at the orifices. Moreover, the hypothesis of laminar flow can be assumed to simplify the modeling. Then:

$$Q_{inlet} = C_i \left(P_{supply} - P(x_h, z_h)\right) \left[\frac{m^3}{s}\right] \qquad (13)$$

where C_i is a coefficient that includes the orifice area and flow coefficient. $P(x_h, y_h)$ is the pressure of the oil at the hole location and P_{supply} is the feeding pressure far from the orifice. A more detailed description of the boundary condition can be found in [30].

In general, a backflow happens if $P_{supply} < P(x_h, z_h)$. Therefore, some oil exits the lands of the SFD and enters in the supply circuit. In practical applications check valves may be used to limit the effect of pressure waves in the supply circuit and to avoid backflows, [31]. In this case, if the oil pressure at the nodes of the holes is higher than the supply one, no boundary condition is assigned.

Central grooves are often present in many applications as shown in Fig. 6.

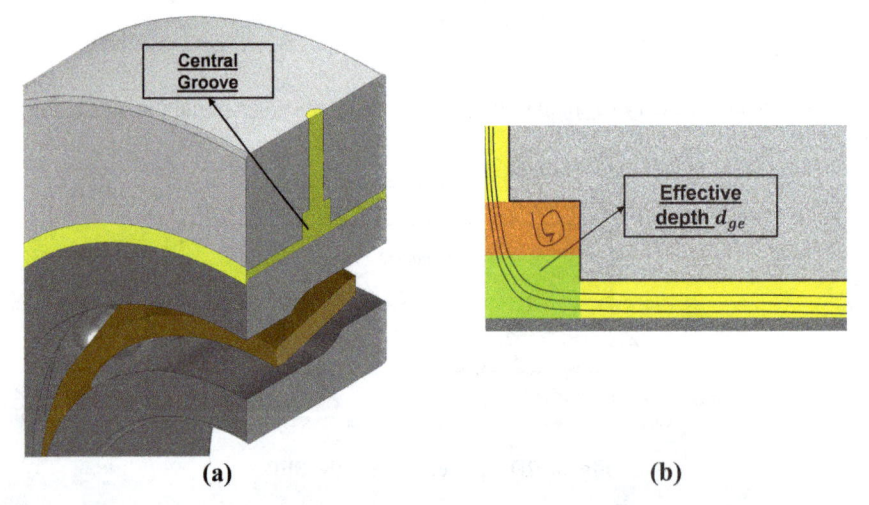

Fig. 6. 3D representation of SFD with central groove (**a**); Schematization of the flow path passing through the central groove for half SFD (**b**).

In older literature, [30, 32], the authors suggest considering the groove as a region where the oil is at the feeding pressure. On the contrary, in [19] the authors report that large levels of dynamic pressure are registered in the groove, proving that the previous assumption was wrong. In this paper, the same approach introduced in [25] is adopted to model the effect of the central groove. The flow inside the groove is considered to be divided into two regions, a recirculating one and through-flow close to the journal, see Fig. 6-b. For the analysis, only the second one is considered effective in the generation of dynamic pressure in the groove. To evaluate the effective groove depth, in [25] the authors characterize experimentally the dynamic coefficients of the SFD. Then, the value of d_{ge} adopted in the model is tuned to obtain the same dynamic characteristics.

Outlet

Different approaches for the outlet boundary condition can be followed, see [5]. When the SFD ends are at ambient pressure the boundary condition is:

$$P(L, t) = P_{air} \tag{14}$$

where $P(L, t)$ is the pressure at the opening if the SFD. In this case the SFD is subjected to high air entrainment, which reduces the dynamic performance of the device. In case of pen ends configuration, larger inlet flow rate of oil is needed to compensate the oil that exits the damper. Therefore, seals are usually applied at the end. The sealing is usually partial to avoid an excessive oil heating that would decrease the SFD damping capacity. Among the various sealing mechanisms, the most common is the piston ring, [5, 31, 33].

The piston ring seal limits the outlet flow rate. It can be modeled as in [30]:

$$q_{out} = \frac{C_p(P(\theta, L) - P_{out})h_p^3}{12\mu w_p} \left[\frac{m^2}{s}\right] \quad (15)$$

where P_{out} is the pressure outside the seal, usually ambient pressure, C_p is the piston ring loss coefficient, $0 < C_p < 1$ and h_p is the piston ring radial gap and w_p is the axial dimension of the piston ring. The C_p coefficient strongly affects the sealing capability. Moreover, it may be difficult to correctly estimate its value. In this work, this coefficient will be considered as a tuning parameter. The authors are aware that the boundary conditions selected for this model represent a simplified approach. More complex and accurate formulations will be evaluated in the future developments.

Circumferential Periodicity

The last boundary condition to be applied corresponds to the circumferential periodicity. To maintain the pressure continuity at the lateral sides of the geometry (Fig. 5), both the pressure and the circumferential pressure gradient along the axial direction must be equal on the two sides. To guarantee that it is sufficient to impose the pressures to be equal at both sides. The resulting circumferential pressure gradient respect the previous hypothesis.

2.7 Forces and Force Coefficients

The Reynolds equation is integrated once the geometry, the mesh, and the boundary conditions are assigned. Eventually the pressure distribution is obtained and is integrated along the circumferential and axial direction to obtain the forces acting on the shaft:

$$\begin{bmatrix} F_x \\ F_y \end{bmatrix} = -\int_0^L \int_0^{2\pi} P(\theta, z, t) \begin{bmatrix} \cos\theta \\ \sin\theta \end{bmatrix} R\, d\theta\, dy \quad (16)$$

The tangential and radial forces, applied on the shaft in the point where the oil thickness is the minimum, can be calculated starting from the forces obtained in the x and y directions with a simple geometrical transformation.

The forces coefficients can be obtained once the forces are known. As reported in many sources, [6] and [19], no stiffness is generated by the damper itself. The damper forces are represented in linearized form:

$$\begin{bmatrix} F_x \\ F_y \end{bmatrix} = -\begin{bmatrix} C_{xx} & C_{xy} \\ C_{yx} & C_{yy} \end{bmatrix}\begin{bmatrix} v_x \\ v_y \end{bmatrix} - \begin{bmatrix} M_{xx} & M_{xy} \\ M_{yx} & M_{yy} \end{bmatrix}\begin{bmatrix} a_x \\ a_y \end{bmatrix} \quad (17)$$

where v_x and v_y and a_x, a_y are the instantaneous journal velocities and accelerations, respectively.

Damping and added mass coefficients along the x and y directions are typical of small shaft displacement around the static equilibrium position. If circular centered orbits are considered, a constant rotating reaction film force is generated by the SFD. In most rotodynamic application, linearized force coefficients are considered. They represent the effect of infinitesimal amplitude motions about the equilibrium position on the bearing reaction forces. These coefficients can be adopted if the previously mentioned hypothesis is valid. However, in SFDs the orbit radius can be comparable to the clearance. The orbit described is far from being close to the equilibrium position. Therefore, the main hypothesis behind linearized force coefficients is violated and an alternative approach should be considered. An orbit-based model, as in [34], is adopted.

3 Model Validation

Numerical and experimental results available in the literature have been considered for the model validation. In [19] the authors tested different geometrical configurations. Therefore, the results reported in [19] have been considered as a reference for this work. Three configurations (SFD B, E and F) have been selected for this paper. The differences among them are related to the clearance, length, presence of end seals and central groove. The diameter of the SFD is 127 mm. Moreover, the oil has density $\rho_L = 805 \text{ kg/m}^3$ and viscosity $\mu_L = 26.5$ mPa.s as in [19]. The characteristics of the SFDs considered for the validation are listed in Table 1.

Table 1. SFDs considered from [19]. d_G and L_G represent the physical depth and length of the central groove, not present in SFD E and F. d_E and L_E represent the depth and length of the grooves at the discharge, not present in SFD E and F. Piston ring seals are applied only for SFD B.

		SFD B	SFD E	SFD F
Length	$L[mm]$	2x 12.7	25.4	25.4
Clearance	$cl[mm]$	0.138	0.122	0.267
Central groove length	$L_G[mm]$	12.5	no	no
Central groove depth	$d_G[mm]$	9.5	no	no
End groove length	$L_E[mm]$	2.5	no	no
End groove depth	$d_E[mm]$	3.5	no	no
Seal	–	yes	no	no

For SFD E and SFD F zero static eccentricity and variable orbit radius are considered. The e/cl ratios considered are 0.05, 0.14, 0.29, and 0.43. The tested frequencies are $10 \div 250$ Hz for SFD E and $10 \div 100$ Hz for SFD F. Since the force coefficients are constant for the selected frequency range, only their values at 100 Hz and 5 0 Hz are shown respectively. SFD B is instead tested in [19] considering a constant obit radius ($e = 0.055cl$) and for different values of static eccentricity. The frequency range considered is 110–210 Hz. Also, for this configuration, the force coefficients remain constant with the frequency. Therefore, the frequency considered for SFD B is 150 Hz. The effective groove depth of the model is tuned to match the force coefficients shown in [19]. The evolution of both the mass and damping coefficients for SFD F is shown in Fig. 7(a, b), for SFD E is shown in Fig. 7(c, d), and for SFD B is shown in Fig. 7(e, f). The results obtained with the current model and the results shown in [19] agree well with each other. For the three configurations, the damping coefficient shows a better agreement between the numerical and the experimental results. On the contrary, the agreement between the numerical and the experimental results for the mass coefficients is lower. This trend may be related to the high values of inlet pressure.

4 Application

A finite beam analysis of a centrifugal compressor has been integrated with the proposed model. The shaft of the machine is 0.7 m long and the nominal diameter is 50 mm. The impeller maximum diameter is 140 mm and the minimum one is 33 mm and the length is 70 mm. The finite element discretization of the structure, with a total of 34 nodes, is shown in Fig. 8.

As shown in Fig. 8, the different elements are different in stiffness and mass diameter. The two roller element bearings supporting the machine are represented by the yellow triangles. Due to the characteristics of the bearings selected the system is barely damped. The sealing element is placed before the impeller and is indicated by the green rectangle. The presence of this component will be considered as the source of the instability. The reduction of the amplitude of the vibration and the mitigation of the instability are studied if the SFD is added to the system.

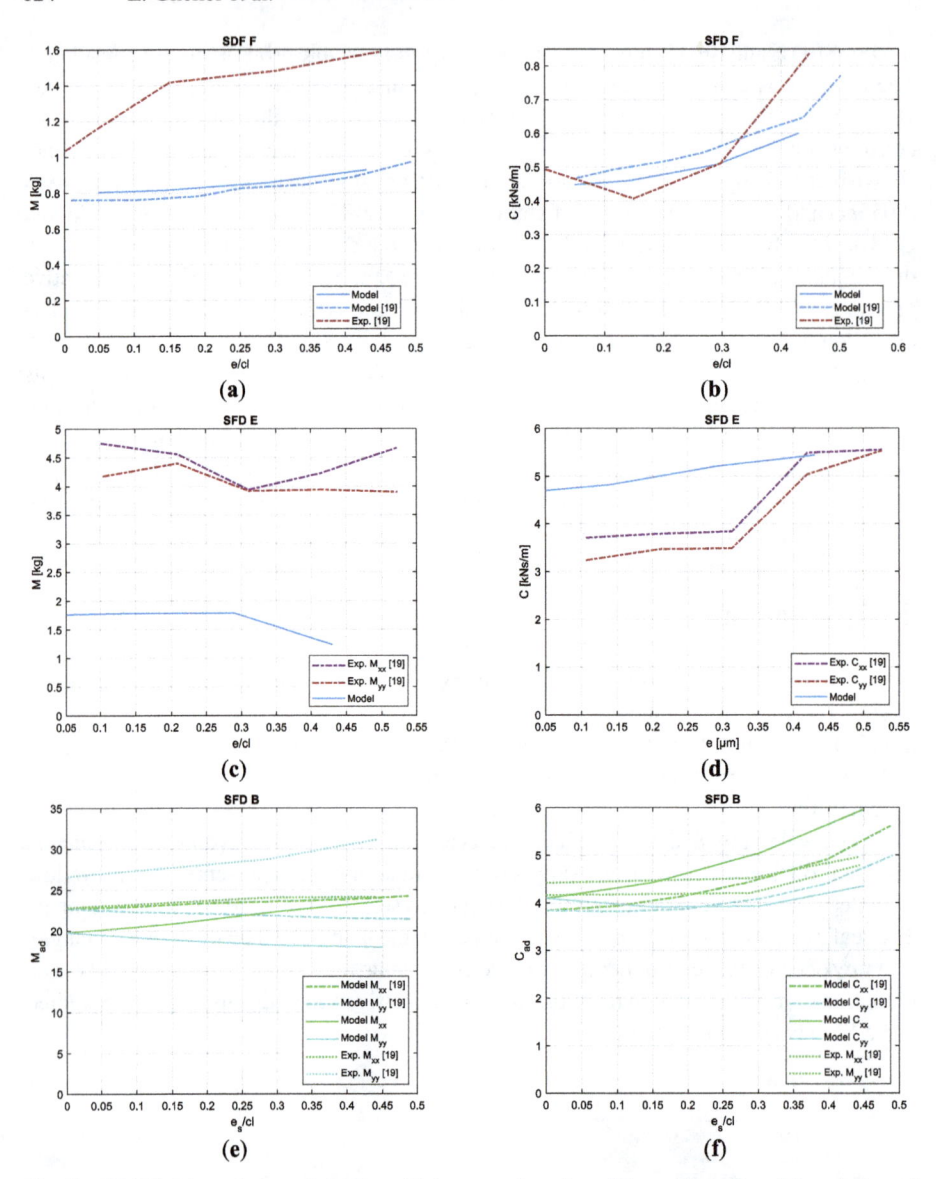

Fig. 7. SFD F: (**a**) evolution of mass coefficient as a function of the orbit radius, (**b**) evolution of damping coefficient as a function of the orbit radius. SFD E: (**c**) evolution of mass coefficient with orbit radius, (**d**) evolution of damping coefficient with orbit radius. SFD B open ends: (**e**) evolution of mass coefficient with static eccentricity, (**f**) evolution of damping coefficient with static eccentricity. Numerical and experimental results from [19].

4.1 Vibration Reduction

An unbalance force of $3 \times 10^{-6}[Kg \cdot m]$ is considered in the analysis at the yellow node of the impeller (node 27). For the vibration reduction analysis, the effect of the sealing

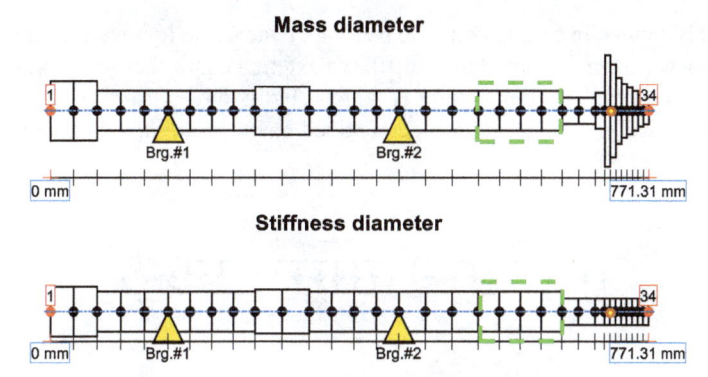

Fig. 8. Mass diameter and stiffness diameter FE discretization of the centrifugal compressor.

element in the green rectangle is not taken into consideration. The focus is placed on the nodes of the impeller because, considering the tight clearances required to maximize the efficiency of the machine, an excessive level of the vibration could be harmful. The operational speed range considered for the compressor goes from 100–300 Hz. 200 Hz is considered as the operating frequency. The forced response to the unbalance at node 27, 29, and 33 of the impeller is shown in Fig. 9. The amplitude of the vibration in the last node at 185 Hz, is close to 2×10^{-4} m. Large vibration amplitudes can compromise the safe operation of the compressor. Considerable damage can result from the contact between the stator and the rotor. For this reason, the level of the vibration, both at the critical speed and at the operating speed, must be reduced as much as possible.

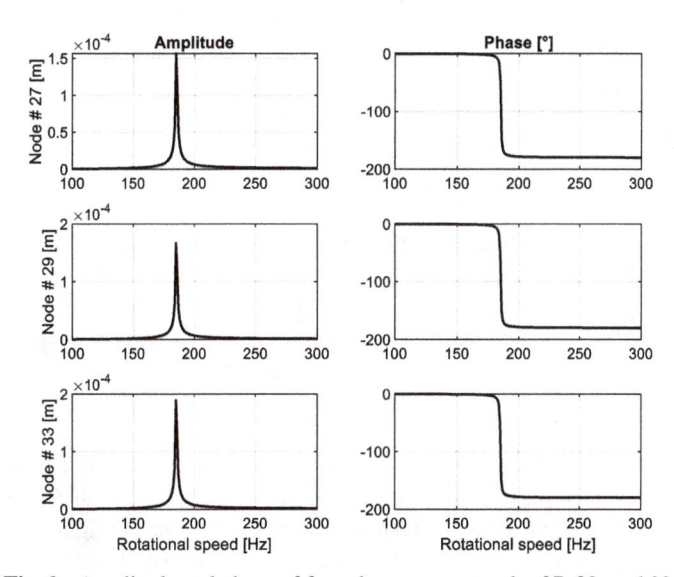

Fig. 9. Amplitude and phase of forced response at nodes 27, 29, and 33.

A SFD is applied in parallel with the bearing on the left to try to reduce the vibration peak. The new system is shown in Fig. 10. An external squirrel cage is supposed to be supporting the SFD. This component is defined by its own stiffness (k_{cage}) and mass (m_{cage}) respectively. An external source of added mass (m_{SFD}) and damping (c_{SFD}) are introduced by the SFD.

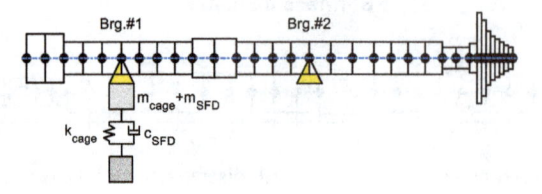

Fig. 10. System discretization with SFD.

A plain SFD without feeding system, seals, and grooves is considered. The characteristics of the SFD and the properties of the ISO VG 46 oil considered are listed in Table 2.

Table 2. Characteristics of the SFD and oil properties considered.

		SFD
Length	$L[mm]$	25
Clearance	$cl[mm]$	0.3
Diameter	$D[mm]$	100
Oil density	$\rho_L [Kg/m^3]$	870
Oil viscosity	$\mu_L [Pa \cdot s]$	0.0775

The forced response of the system when the squirrel cage is added to the system and when the SFD is applied is shown in Fig. 11. The effect that is obtained on the forced response of the system when the squirrel cage is added is the same as if a tuned mass damper is considered. The vibration peak is moved at a lower frequency, but the amplitude is not strongly affected. On the contrary, when the SFD dynamic effect is considered, the system is highly damped, and the amplitude of the vibration peak is strongly reduced.

To assess the effect of the SFD clearance on its dynamic characteristics, a parametric investigation has been performed. Considering different values of SFD clearance, the ratio between the peak of the force response at node 33 in the configuration with SFD and the original configuration without SFD, is shown in Fig. 12. If the ratio is lower than one, then the addition of the SFD determines a reduction of the vibration amplitude. The SFDs with the highest clearance minimize the forced response at the operating frequency. When the SFD clearance is 0.1 mm, the amplitude of the vibration is almost identical as

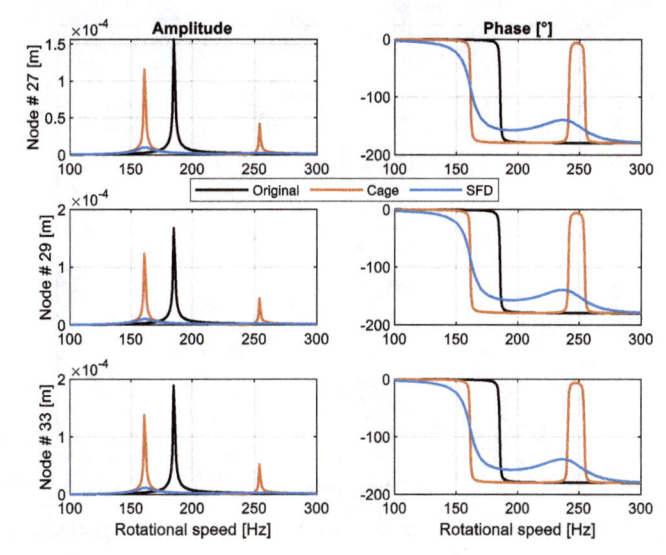

Fig. 11. Forced response comparison for original configuration, configuration with squirrel cage application and configuration with SFD.

the one of the original configuration. The evolution of the force coefficients of the SFD with the clearance is shown in Fig. 13. It is possible to appreciate that the reduction of the SFD clearance determines an increase in the force coefficients. However, the increase of the force coefficients does not determine a reduction of the vibration amplitude at the operation frequency considered for this application.

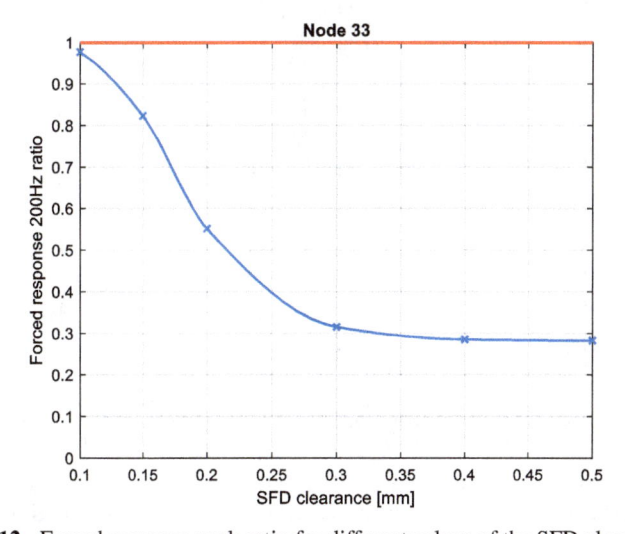

Fig. 12. Forced response peak ratio for different values of the SFD clearance.

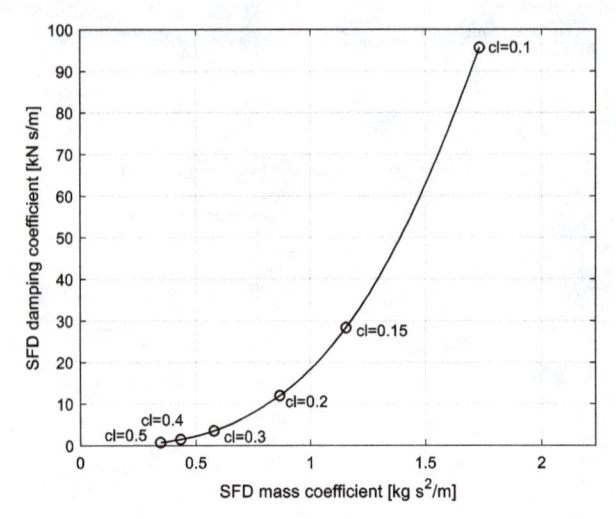

Fig. 13. Evolution of force coefficients at 200 Hz with the SFD clearance.

The evolution of the forced response at node 33 with the SFD clearance is shown in Fig. 14. The SFD that minimizes the forced response at 200 Hz is not the one that guarantees the minimization of the first amplitude peak. Moreover, the SFD clearance seems to have no impact on the second peak around 250 Hz.

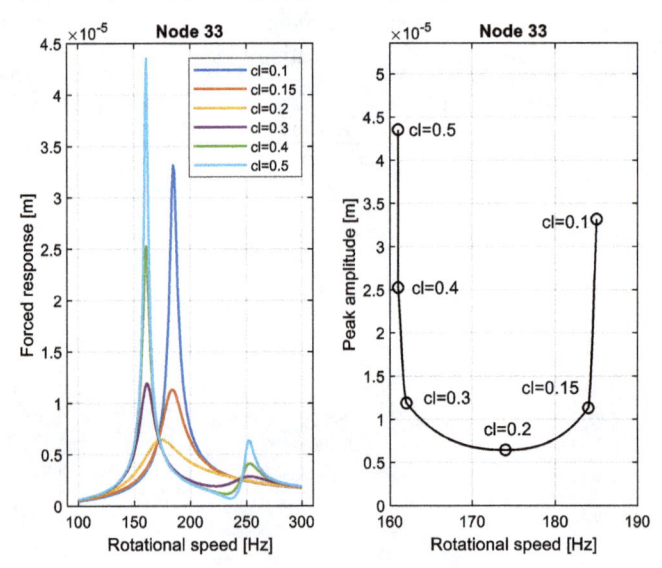

Fig. 14. Evolution of the forced response of the system at node 33 for the different SFD clearances considered and evolution of peak amplitude and first critical speed with SFD clearance.

The complexity of selecting the most appropriate SFD configuration is highlighted from this analysis. This decision is strongly dependent on the application that is considered. Moreover, the selection of the SFD configuration is strongly affected by the goal that must be obtained (peak minimization or amplitude reduction at operational frequency).

4.2 Correction of Instability

In this section the effect of the seal placed before the impeller is considered as source of instability. The stiffness matrix at the nodes of the seal is introduced as follows:

$$K_{seal} = \begin{bmatrix} 0 & k_{seal} \\ -k_{seal} & 0 \end{bmatrix} \tag{18}$$

The value of k_{seal} is varied to establish when the sealing presence destabilizes the system resulting in positive values of the real parts of the eigenvalues of the system. The evolution of the maximum real part of the eigenvalues of the system with the increase of k_{seal} is shown in Fig. 15. The first instability is encountered at $k_{seal} = 15000\,\text{N/m}$. However, it is at $k_{seal} = 17500\,\text{N/m}$ that the system is unstable for the whole speed range. It is worth to notice that until $k_{seal} = 35000\,\text{N/m}$ only the real part of the first eigenvalue assumes positive values, for higher levels of k_{seal} also the real part of the fifth eigenvalue is positive.

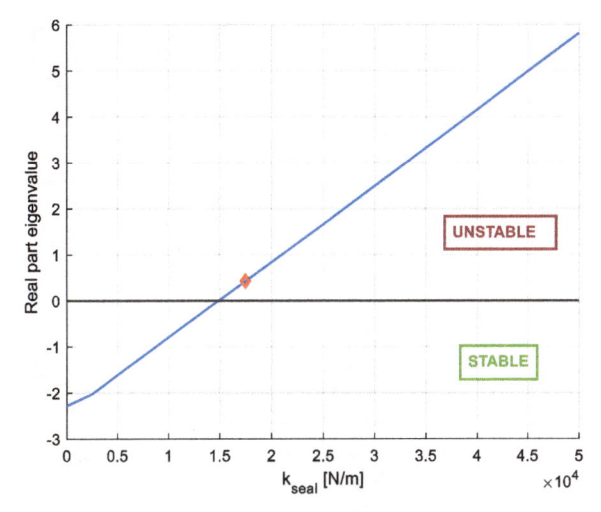

Fig. 15. Evolution of maximum real part of eigenvalues with k_{seal}.

For this analysis it is considered $k_{seal} = 17500\,\text{N/m}$ since it is the first value at which the system is unstable for the whole speed range. To limit the instability, a SFD is applied in parallel with the first bearing as shown in Fig. 10. The addition of the SFD to the system will introduce damping that will counteract the presence of the instability.

To study the stabilizing effect of the SFD, the evolution of the dimensionless damping factor with the speed is considered. This indicator is defined as follows:

$$\eta_i = -\frac{Real(\lambda_i)}{Imag(\lambda_i)} \tag{19}$$

where λ_i is the eigenvalue and η_i is the dimensionless damping factor for the i-th node. When η_i is positive the relative mode is unstable.

The tested SFD has the same characteristics of the one shown described in Table 2 but the clearance is set to 0.5 mm. The force coefficients obtained are like the one shown in Fig. 13. The evolution of the dimensionless damping factor for the original system and the SFD system for the unstable mode is shown in Fig. 16. As expected, for the original system, the dimensionless damping factor is negative for the whole speed range. The same holds for the configuration with the addition of the cage. On the contrary, the addition of the SFD can stabilize the system for the whole frequency range considered.

Then a parametric analysis on k_{seal} has been performed to investigate at which extent the application of the SFD is able to stabilize the system. The evolution of the dimensionless damping factor for different values of k_{seal} for the system with the SFD is shown in Fig. 17. For the SFD configuration considered the system is stabilized until $k_{seal} = 5e4 \, \text{N/m}$. In this case, also the 7th mode of the original system results unstable. On the contrary, when the SFD is applied, only the 1st mode is unstable. In case of higher values of k_{seal}, a different SFD configuration may guarantee the stability of the system.

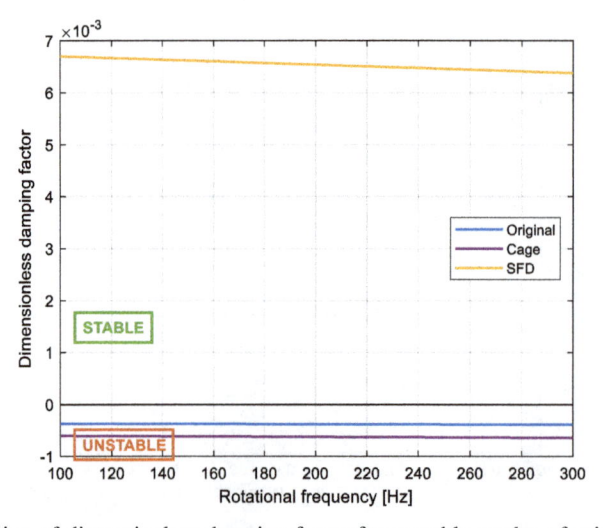

Fig. 16. Evolution of dimensionless damping factor for unstable modes of original, cage, and SFD system with rotational frequency.

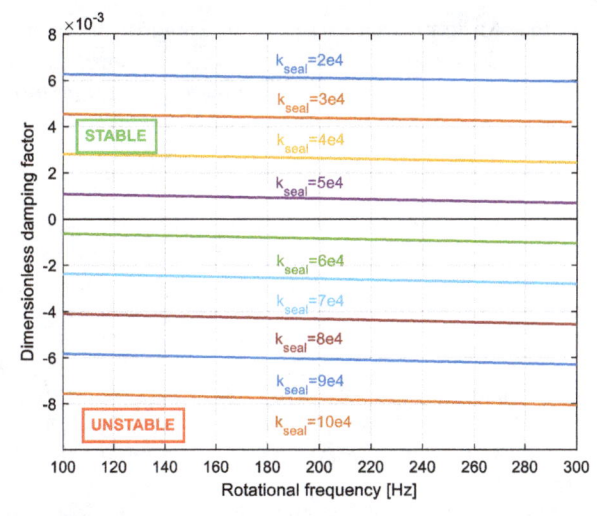

Fig. 17. Evolution of dimensionless damping factor for the unstable mode for different values of k_{seal} for the system with the SFD.

5 Conclusions

The state of the art on SFDs is investigated in this paper. The characteristic features of these components are highlighted and a review on the modeling strategies is given. A comprehensive model based on the 2D Reynolds equation is presented in the details. The classical Reynolds equation is modified so to include the modeling of both the air entrainment and the temporal inertia.

The model is validated with both numerical and experimental results available in the literature. Then the model is integrated into a finite element code developed to study the dynamic response of turbomachines. The effectiveness of the SFD in the reduction of the level of the vibrations is investigated. From the results shown it is evident that the introduction of the SFD can drastically reduce the amplitude of the vibration. However, the optimal configuration to be selected strongly depends on the application considered and on the wanted effect. For example, it is shown that the optimal configuration to reduce the level of the vibration at a given frequency is not the same that minimize the amplitude of the vibration peak.

Secondly, the effectiveness of SFDs to mitigate the presence of destabilizing phenomena is investigated. The same turbomachine considered for the previous analysis is considered and the presence of a destabilizing sealing element is added. Different levels of source instability have been considered. The results obtained show that SFD are, in general, able to solve the presence of the instability. Also in this case, a trade-off may be needed when designing the SFD if both the vibration reduction and the instability correction are wanted.

The model proposed can be considered predict effectively and efficiently the dynamic behavior of SFDs. Moreover, it can be easily integrated in models used for the rotor dynamic analysis of turbomachines. The authors are aware that more accurate and precise models are present in the literature. However, some of them are more complex and require higher simulation time. An improvement in the modeling of the boundary conditions and in general in the modeling of the other characterizing aspects related to the SFDs may be required when studying more complex geometrical configurations.

References

1. Zeidan, F.Y., San Andres, L., Vance, J.M.: Design and application of squeeze film dampers in rotating machinery. In: Proceedings of the 25th Turbomachinery Symposium, pp 169–188. Texas A&M University (1996)
2. Vance, J.M., Zeidan, F.Y., Murphy, B.: Machinery Vibrations and Rotordynamics. Wiley, Hoboken (2010)
3. Gunter Jr., E.J., Barrett, L.E., Allaire, P.E.: Design and application of squeeze film dampers for turbomachinery stabilization, pp. 127–141 (1975). https://doi.org/10.21423/R1T37D
4. Chen, W.J., Rajan, M., Rajan, S.D., Nelson, H.D.: Optional design of squeeze ilm dampers for flexible rotor systems. J. Mech. Transm. Autom. Design 110, 166–174 (1988). https://doi.org/10.1115/1.3258922
5. della Pietra, L., Adiletta, G.: The squeeze film damper over four decades of investigations. Part I: characteristics and operating features. Shock Vibr. Digest 34, 3–26 (2002)
6. San Andrés, L.: Squeeze film damper: operation, models and theoretical issues. In: Modern Lubrication Theory. Texas A&M University (2012)
7. San Andrés, L., Vance, J.M.: Effects of fluid inertia and turbulence on the force coefficients for squeeze film dampers. J. Eng. Gas Turbines Power 108, 332–339 (1986). https://doi.org/10.1115/1.3239908
8. San Andrés, L.A., Vance, J.M.: Effects of fluid inertia on finite-length squeeze-film dampers. ASLE Trans. 30, 384–393 (1987). https://doi.org/10.1080/05698198708981771
9. Hamzehlouia, S., Behdinan, K.: Squeeze film dampers executing small amplitude circular-centered orbits in high-speed turbomachinery. Int. J. Aerosp. Eng. 2016 (2016). https://doi.org/10.1155/2016/5127096
10. Zeidan, F.Y., Vance, J.M.: Cavitation regimes in squeeze film dampers and their effect on the pressure distribution. Tribol. Trans. 33, 447–453 (1990). https://doi.org/10.1080/10402009008981975
11. Diaz, S.E., San Andrés, L.A.: Air entrainment vs. lubricant vaporization in squeeze film dampers: an experimental assessment of their fundamental differences. In: Proceedings of the ASME Turbo Expo (1999)
12. Elrod, H.G.: Cavitation algorithm. J. Lubr. Technol. 103, 350–354 (1981). https://doi.org/10.1115/1.3251669
13. Diaz, S.E., Andres, L.A.: Reduction of the dynamic load capacity in a squeeze film damper operating with a bubbly lubricants. J. Eng. Gas Turbines Power 121, 703–709 (1999). https://doi.org/10.1115/1.2818530
14. San Andrés, L., Diaz, S.E.: Flow visualization and forces from a squeeze film damper operating with natural air entrainment. J. Tribol. 125, 325–333 (2003). https://doi.org/10.1115/1.1510878
15. Diaz, S.E.: An engineering model for prediction of forces in SFD's and experimental validation for operation with entrainment. Texas A&M University (1999)
16. Diaz, S., San Andrés, L.: A model for squeeze film dampers operating with air entrainment and validation with experiments. J. Tribol. 123, 125–133 (2001). https://doi.org/10.1115/1.1330742
17. Méndez, T.H., Torres, J.E., Ciaccia, M.A., Díaz, S.E.: On the numerical prediction of finite length squeeze film dampers performance with free air entrainment. J. Eng. Gas Turbines Power 132 (2010). https://doi.org/10.1115/1.2981182

18. Gehannin, J., Arghir, M., Bonneau, O.: Evaluation of Rayleigh-Plesset equation based cavitation models for squeeze film dampers. J. Tribol. **131**, 1–4 (2009). https://doi.org/10.1115/1.3063819
19. San Andrés, L., Jeung, S., Den, S., Savela, G.: Squeeze film dampers: an experimental appraisal of their dynamic performance. In: Proceedings of Asia Turbomachinery Pump Symposium. Singapore (2016)
20. Gehannin, J., Arghir, M., Bonneau, O.: Complete squeeze-film damper analysis based on the "Bulk Flow" equations. Tribol. Trans. **53**, 84–96 (2010). https://doi.org/10.1080/10402000903226382
21. Xing, C., Braun, M.J., Li, H.: A three-dimensional navier-stokes-based numerical model for squeeze-film dampers. Part 1-effects of gaseous cavitation on pressure distribution and damping coefficients without consideration of inertia. Tribol. Trans. **52**, 680–694 (2009). https://doi.org/10.1080/10402000902913303
22. Xing, C., Braun, M.J., Li, H.: A three-dimensional Navier-Stokes-based numerical model for squeeze film dampers. Part 2-effects of gaseous cavitation on the behavior of the squeeze film damper. Tribol. Trans. **52**, 695–705 (2009). https://doi.org/10.1080/10402000902913311
23. Zhou, H.-L., Chen, X., Zhang, Y.-Q., et al.: An analysis on the influence of air ingestion on vibration damping properties of squeeze film dampers. Tribol. Int. **145** (2020). https://doi.org/10.1016/j.triboint.2020.106168
24. Lee, G.J., Kim, J., Steen, T.: Application of computational fluid dynamics simulation to squeeze film damper analysis. J. Eng. Gas Turbines Power **139** (2017). https://doi.org/10.1115/1.4036511
25. Delgado, A., San Andrés, L.: A model for improved prediction of force coefficients in grooved squeeze film dampers and oil seal rings. J. Tribol. **132**, 1–12 (2010). https://doi.org/10.1115/1.4001459
26. Fan, T., Hamzehlouia, S., Behdinan, K.: The effect of lubricant inertia on fluid cavitation for high-speed squeeze film dampers. J. Vibroeng. **19**, 6122–6134 (2017). https://doi.org/10.21595/jve.2017.19314
27. Giacopini, M., Fowell, M.T., Dini, D., Strozzi, A.: A mass-conserving complementarity formulation to study lubricant films in the presence of cavitation. J. Tribol. **132** (2010). https://doi.org/10.1115/1.4002215
28. Almqvist, A., Fabricius, J., Larsson, R., Wall, P.: A new approach for studying cavitation in lubrication. J. Tribol. **136** (2014). https://doi.org/10.1115/1.4025875
29. Hamzehlouia, S., Behdinan, K.: A study of lubricant inertia effects for squeeze film dampers incorporated into high-speed turbomachinery. Lubricants **5** (2017). https://doi.org/10.3390/lubricants5040043
30. Marmol, R.A., Vance, J.M.: Squeeze film damper characteristics for gas turbine engines. J. Mech. Design Trans. ASME **100**, 139–146 (1978). https://doi.org/10.1115/1.3453878
31. San Andrés, L.S., Koo, B., Jeung, S.-H.: Experimental force coefficients for two sealed ends squeeze film dampers (piston rings and O-rings): an assessment of their similarities and differences. J. Eng. Gas Turbines Power **141** (2019). https://doi.org/10.1115/1.4040902
32. Lund, J.W., Smalley, A.J., Tecza, A.J., Walton, J.F.: Squeeze-film damper technology: part 1 - prediction of finite length damper performance (1983). https://doi.org/10.1115/83-gt-247
33. San Andrés, L., Koo, B.: Effect of lubricant supply pressure on SFD performance: ends sealed with O-rings and piston rings. In: Cavalca, K.L., Weber, H.I. (eds.) IFToMM 2018. MMS, vol. 60, pp. 359–371. Springer, Cham (2019). https://doi.org/10.1007/978-3-319-99262-4_26
34. San Andrés, L., Jeung, S.-H.: Orbit-model force coefficients for fluid film bearings: a step beyond linearization. J. Eng. Gas Turbines Power **138** (2016). https://doi.org/10.1115/1.4031237

Investigation of Active Configuration in Gas Foil Bearings for Stable Ultra High-Speed Operation

Ioannis Gavalas, Anastasios Papadopoulos, and Athanasios Chasalevris[✉]

National Technical University of Athens, 15780 Athens, Greece
chasalevris@mail.ntua.gr

Abstract. Gas foil bearings (GFBs) are machine elements used to support rotating shafts in high-speed applications, receiving increasing interest due to their oil-free feature (gas lubrication), reliability and simplicity. However, their applicability is limited due to the low load capacity as the dynamic viscosity of the gas (ambient air, or other) is lower than this of the oil (or other fluid). In the last 50 years, gas foil bearings have increased (at least doubled) their load capacity as pioneering work on the bearing design has taken place; this was mainly on tribological aspects of the foil surface and on dynamic aspects of foil properties. GFBs still face instability issues at high speeds, at least at higher speeds than the respective threshold speed of instability of an oil lubricated bearing. This paper investigates the potential to increase stability threshold of GFBs further, by active configuration of the foil shape, in order to render the optimum stability characteristics (higher damping) at discrete speeds. The analysis includes a simple rigid Jeffcott rotor model with unbalance, mounted on two active gas-foil bearings (AGFBs). The gas lubrication problem is coupled to thermal flow and structural deformation of the foil. An optimization technique is used to configure the foil according to a stability index, this being the dominant pair of eigenvalues. It is found that specific foil configurations can establish an instability-free operating range up to DN values (DN = Diameter [mm] times N [RPM]) corresponding to the speed of sound (DN ≈ 6.5e6), for both small (D30) and large (D100) AGFB applications.

Keywords: active gas bearings · rotating shafts · nonlinear dynamics · stability

1 Introduction

Gas foil bearings (GFBs) are the primary emerging technology for the transition to oil-free ultra high-speed rotating machinery. Compared to traditional oil lubricated or rolling element bearings, GFBs are able to operate in high rotating speeds and very high temperatures with essentially no maintenance and increased durability and efficiency.

In contrast to these advantages, GFBs generally suffer from strong gas film-induced instabilities which are characterized by high amplitude subsychronous vibrations, commonly associated with the oil film equivalent terms: oil whirl and oil whip. Another strong drawback is the lower load capacity compared to equivalent oil film bearings which is increased by the compliant foil structure.

There have been numerous studies and an increasing interest of modifying the surface shape of the top foil of the GFB to increase the threshold speed of instability. Kim and San Andres [1] first introduced three metal shims to the bearing structure to examine the effect of mechanical preloading on the rotordynamics of GFBs. It was shown that mechanical preloading reduces the difference between the cross coupling coefficients of the gas film which leads to an increased onset speed of instability. Later, Sim et al. [2] revealed that an optimum preload and bearing clearance combination exists in connection to the onset speed of instability and power loss, Schiffman et al. [3] showed an optimum preload pattern which should be dependent on the design characteristics of the GFB and Walter [4] demonstrated that a an optimum preload value exists with respect to stability, load capacity and lift-off speed.

Mechanical preloading is based on the principle of increasing the wedge length of the gas film in order to increase the pressure at the preload locations which leads to increased direct stiffness and damping coefficients and consequently to the increase of the onset speed of instability.

This method, however simple and effective, has some important limitations. Mechanical preloading reduces the gas film thickness and therefore bearing performance is reduced at lower operating speeds (load capacity is decreased and lift-off speed is increased) [5] while bearing wear is increased [6,7]. Also the GFB has a slower response to variations in operation conditions such as rotor speed and external excitation [8].

In recent years, several works addressed these limitations by embedding piezoelectric (PZT) actuators to the bearing structure which are usually referred to as Active or Controllable Gas Foil Bearings (AGFBs, C-GFBs).

Sadri et al. [5,9] and Feng et al. [6] implemented patch-type PZT actuators used to deform the outer housing shell of the bearing structure in order to control the bore shape in different regions of operating speeds and adapting to the relevant performance requirements. Park and Sim [10] implemented a C-GFB with 9 PZT actuators that deform the bearing housing and support in order to control mechanical preload and bearing clearance and influence the stable passage through critical speeds and the amplitude of vibrations. It was also demonstrated that bearing clearance control has a significant effect on the direct stiffness and damping coefficients and mechanical preload control has a large influence on the cross-coupling effect. Finally, Park et al. [8] performed linear and nonlinear rotordynamic analysis and experimental characterization with parameter identification of a C-GFB with three PZT actuators using preload control and demonstrated a significant increase of the OSI and Guan et al. [7] applied the bearing concept of Feng et al. [6] to demonstrate open loop performance and bifurcations and to perform real-time control of rotor vibrations using controllable mechanical preload with PID controller structure.

The objective of this work is to investigate the performance characteristics of a novel AGFB where the bump foil and therefore the dry friction induced damping is absent and the compliant top foil is mounted on a number of PZT actuators both in the circumferential and axial direction with linear stiffness and damping properties and mounted on the bearing shell, assumed rigid. A predictive Thermo-Elasto-Aerodynamic (TEAD) model is considered and used to perform nonlinear analyses and a design optimization scheme is used to calculate optimum PZT actuator displacements to establish stable operation.

2 Modelling System Dynamics

In this section, an overview of the nonlinear dynamic model is presented. The physical model of the rotor-AGFB system consists of a rigid symmetric rotor mounted on two identical AGFBs and carrying a disc at the bearing mid-span, as shown in Fig. 1a. The AGFBs have a compliant top foil structure which is pin supported on a number of linear actuators along the circumferential and axial direction, see Fig. 1a, 1b and the other end of the linear actuators is fixed on the rigid bearing shell. These actuators replace the bump foil structure in order to eliminate the damping induced to the system due to bump foil related material hysteresis and dry friction and shift the focus to the gas film. The linear actuators at a later stage of research will be substituted by more realistic piezoelectric stacks with an additional layer of a material with controllable stiffness and damping properties (e.g. electro-magnetorheological polymers). By applying a desired displacement on the free end of the actuators the flexible top foil structure can be deformed to a large variety of configurations that can potentially further increase the stability threshold. The journal and the compliant top foil structure are coupled via the resultant gas film forces shown in Fig. 1c.

2.1 Rigid Rotor on Active Gas Foil Bearings

The equations of motion for the perfectly aligned rigid rotor are defined in Eq. (1) for lateral motion

$$\ddot{x}_j = \frac{1}{m_j}(F_{b,x} + F_{u,x}), \quad \ddot{y}_j = \frac{1}{m_j}(F_{b,y} + F_{u,y} + F_g) \qquad (1)$$

where x_j, y_j are the rigid rotor center (journal) displacements in the horizontal and vertical direction, $m_j = M/2$ is the journal mass, $F_{b,x}, F_{b,y}$ are the resulting gas film forces in the horizontal and vertical direction (evaluation shown in Sect. 2.2), $F_{u,x} = U\Omega^2 cos(\Omega t), F_{u,y} = U\Omega^2 sin(\Omega t)$ are the unbalance forces at a constant operating speed Ω, $U = 10^{-3}G/\Omega_r$ is the unbalance in [kgm] as defined by ISO 21940-11 [11] for rated speed Ω_r and balancing grade G, and $F_g = m_j g$ is the force due to the gravitational acceleration $g = 9.81$ m/s^2.

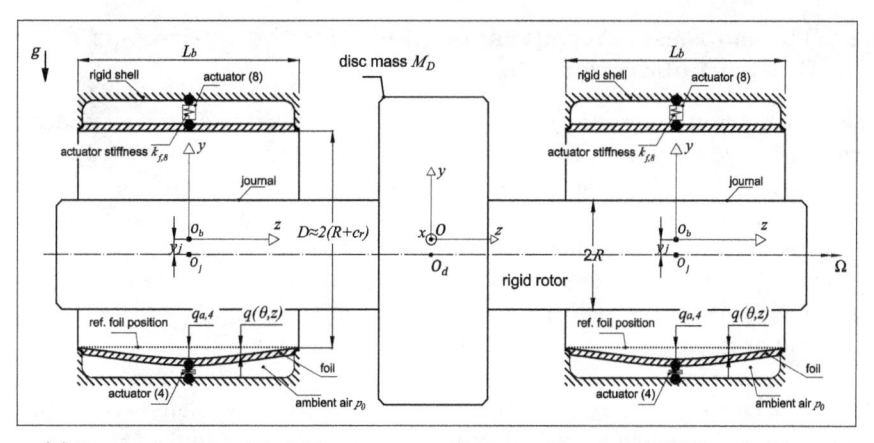

(a) Representation of a rigid rotor mounted on two identical AGFBs carrying a mass MD in the bearing midspan. AGFB actuators perform identical motions at both bearings.

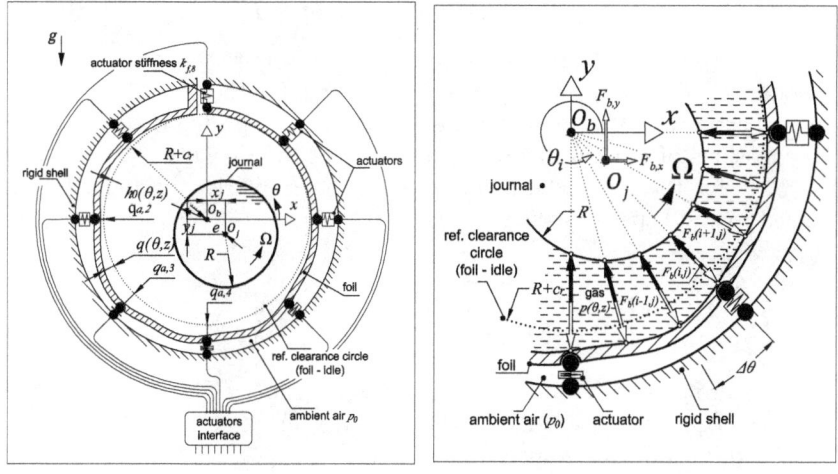

(b) Representation of AGFB front view, key geometry and operating parameters.

(c) Detail of gas pressures and forces acting on the rotating journal and compliant foil structure.

Fig. 1. Representation of the investigated model of a rigid rotor mounted on AGFBs

The ODEs in Eq. (1) can be rearranged in state space representation using the rotor state variable $\boldsymbol{q}_r = [x_j \ y_j \ \dot{x}_j \ \dot{y}_j]^\mathsf{T}$ and shown in Eq. (2).

$$\dot{\boldsymbol{q}}_{r4\times 1} = \begin{bmatrix} \mathbf{0} & \mathbf{I} \\ \mathbf{0} & \mathbf{0} \end{bmatrix} \boldsymbol{q} + \begin{Bmatrix} \mathbf{0} \\ 1/m_j(F_{b,x} + F_{u,x}) \\ 1/m_j(F_{b,y} + F_{u,y}) \end{Bmatrix} \quad (2)$$

2.2 Thermo-Elasto-Aerodynamic Model for the Active Gas Foil Bearing Lubrication

The AGFB top foil is modeled using the well known plate element by Merosh, Zienkiewicz and Cheung (MZC) with analytical integration to calculate the stiffness matrix $\mathbf{K}$ and mass matrix $\mathbf{M}$, while a small value of $\delta = 10^{-5}$ is considered for stiffness-proportional damping. Then, the damping matrix will be equal to $\mathbf{C}_f = \delta \mathbf{K}_f$. The number of intervals in the orthogonal Finite Element mesh is equal to $N_x \times N_z$, and the number of nodes is equal to $(N_x + 1)(N_z + 1)$. The applied fixed constraint reduces the number of nodes to $N_x(N_z + 1)$ and the size of the full order stiffness, damping and mass matrices is equal to $3N_x(N_z + 1)$.

The mass, damping and stiffness matrix of the foil FE model are then reduced using static (Guyan) reduction retaining only the transverse displacement degrees of freedom q_i and omitting the angular displacements $\theta_{x,i}, \theta_{y,i}$, and then the reduced order equations of motion for the foil can be written in Eq. (3).

$$\mathbf{M}_{f,r}\ddot{\mathbf{q}} + \mathbf{C}_{f,r}\dot{\mathbf{q}} + \mathbf{K}_{f,r}\mathbf{q} = \mathbf{F}_p + \mathbf{F}_a \tag{3}$$

where $F_{p,i} = (p_i - p_\infty)\Delta x \Delta z$ is the normal force due to gas film pressure at every node, and $F_{a,j} = c_{a,j}(\dot{q}_j - \dot{q}_{a,j}) + k_{a,j}(q_j - q_{a,j})$ is the normal force due to the applied actuator displacement $q_{a,j}$ at the foil pin supported (actuator) nodes j. The equations of motion in Eq. (3) are solved in the context of the coupled ODE system for the foil nodal displacements q_f. Similarly, the foil ODEs are rearranged to state space as shown in Eq. (4).

$$\begin{Bmatrix} \dot{\mathbf{q}}_{(N_x+1)N_z \times 1} \\ \ddot{\mathbf{q}}_{(N_x+1)N_z \times 1} \end{Bmatrix} = \begin{bmatrix} \mathbf{0} & \mathbf{I} \\ -\mathbf{M}_{f,r}^{-1}\mathbf{K}_{f,r} & -\mathbf{M}_{f,r}^{-1}\mathbf{C}_{f,r} \end{bmatrix} + \begin{Bmatrix} \mathbf{0} \\ \mathbf{M}_{f,r}^{-1}(\mathbf{F}_p + \mathbf{F}_a) \end{Bmatrix} \tag{4}$$

The pressure p, temperature T and density ρ of the gas film are described by the Reynolds equation coupled with the thin film energy equation and the gas law (assumed ideal in this study) which are given in Eqs. (5), (6) and (7), [12].

$$\frac{\partial}{\partial t}(\rho h) + \frac{\Omega}{2}\frac{\partial}{\partial \theta}(\rho h) = \frac{1}{R^2}\frac{\partial}{\partial \theta}\left(\frac{\rho h^3}{12\mu}\frac{\partial p}{\partial \theta}\right) + \frac{\partial}{\partial z}\left(\frac{\rho h^3}{12\mu}\frac{\partial p}{\partial z}\right) \tag{5}$$

$$\rho c_v h \left[\frac{\partial T}{\partial t} + \frac{\Omega}{2}\frac{\partial T}{\partial \theta} - \frac{h^2}{12\mu}\left(\frac{1}{R^2}\frac{\partial p}{\partial \theta}\frac{\partial T}{\partial \theta} + \frac{\partial p}{\partial z}\frac{\partial T}{\partial z}\right)\right]$$
$$= -\alpha_v T p \left[\frac{\partial h}{\partial t} + \frac{\Omega}{2}\frac{\partial h}{\partial \theta} - \frac{1}{R^2}\frac{\partial}{\partial \theta}\left(\frac{h^3}{12\mu}\frac{\partial p}{\partial \theta}\right) - \frac{\partial}{\partial z}\left(\frac{h^3}{12\mu}\frac{\partial p}{\partial z}\right)\right] \tag{6}$$
$$-\beta(T - T_\infty) + R^2\Omega^2\frac{\mu}{h} + \frac{h^3}{12\mu}\left[\frac{1}{R^2}\left(\frac{\partial p}{\partial \theta}\right)^2 + \left(\frac{\partial p}{\partial z}\right)^2\right]$$

$$p = \rho R_s T \tag{7}$$

where μ is the dynamic viscosity of the gas which is assumed to follow Sutherland's Law [13], given in Eq. (8), β is an empirical heat transfer coefficient,

$$\frac{\mu}{\mu_0} = \left(\frac{T}{T_0}\right)^{3/2} \frac{T_0 + S_\mu}{T + S_\mu} \qquad (8)$$

$a_v = 1/p\, \partial p/\partial T$ is the compressibility coefficient, c_v is the specific heat (assumed constant for ideal gas) and R_s is the specific ideal gas constant [14]. Finally, T_∞ is the ambient temperature. The Reynolds (5) and thin film energy equation (6) are discretized in the spatial coordinates θ, z with finite differences and converted to two systems of ODEs by selecting the density ρ and temperature T as the primary variables, see Eq. (9). It is noted that the finite element foil nodes and the finite difference grid points are collocated but that should not be necessary the case provided suitable interpolation functions are available. The boundary conditions at the bearing inlets (at $\theta = \pi/2$ and the cavitation region) are $p = p_\infty, T = T_\infty$ and at the bearing outlets, at $\theta = 5\pi/2$ and the region where $\partial p/\partial z(\theta, z = 0) > 0$ (the inequality is reversed at $z = L_b$), are $p = p_\infty$ while the temperature T is extrapolated from the computational domain. Furthermore, Gümbel's boundary condition is used which effectively discards subambient pressures when calculating the forces acting on the foil and the journal.

$$\begin{aligned}\dot{\rho}_{N_x N_z \times ,1} &= \boldsymbol{f}_{Re}(\boldsymbol{\rho}, \boldsymbol{T}, \boldsymbol{p}, \boldsymbol{h}, \varOmega, \mu), \\ \dot{\boldsymbol{T}}_{N_x N_z \times ,1} &= \boldsymbol{f}_E(\boldsymbol{\rho}, \boldsymbol{T}, \boldsymbol{p}, \boldsymbol{h}, \varOmega, \mu, c_v, a_v, \beta)\end{aligned} \qquad (9)$$

The equations of gas film flow, energy and state and the corresponding three unknown state variables of pressure, density and temperature are coupled to the equations of motion of the top foil through the nodal forces due to the gas film pressure $\boldsymbol{F}_p$ and the equations of motion of the journal through the film thickness h given in Eq. (10) for a perfectly aligned journal and the resultant bearing forces given in Eq. (11) where the integrals are evaluated by the trapezoidal method.

$$h_i = c_r - x_j \cos(\theta) - y_j \sin(\theta) + q_i \qquad (10)$$

$$\begin{aligned}F_{b,x} &= \int_0^{2\pi} \int_0^{L_b} (p(\theta, z) - p_\infty)\cos(\theta) dz d\theta \\ F_{b,y} &= \int_0^{2\pi} \int_0^{L_b} (p(\theta, z) - p_\infty)\sin(\theta) dz d\theta\end{aligned} \qquad (11)$$

The final unknown state vector is then $\boldsymbol{x}_s = [\boldsymbol{\rho}^\mathsf{T}\ \boldsymbol{T}^\mathsf{T}\ \boldsymbol{q}_f^\mathsf{T}\ \boldsymbol{q}_r^\mathsf{T}]^\mathsf{T}$ and the coupled ODE system of $2(N_x N_z) + 2N_x(N_z+1) + 4$ parameterized for the rotating speed $\varOmega$ is defined in Eq. (12).

$$\dot{\boldsymbol{x}}_s = \boldsymbol{f}(\boldsymbol{x}_s, \varOmega, t) \qquad (12)$$

3 Evaluation of Response of the Dynamic System

The methods for solving for fixed point and limit cycle equilibria, as well as for time transient oscillations, are outlined in the first part of this section. Next the optimization procedure for load capacity and stability is established.

3.1 Methods for Nonlinear Dynamic Analysis

The evaluation of fixed points (equilibria) is performed using the well-known Broyden's "bad" method using a complex step derivation formula for the evaluation of the Jacobian.

Broyden's method has the core advantage of requiring the evaluation of the Jacobian, a difficult and expensive operation, only for the first iteration and performing linear updates for the remaining iterations. Broyden's "bad" method has the additional advantage of updating the inverse Jacobian directly thus skipping the expensive operation of inverting a large matrix. The main disadvantage of Broyden's methods (and quasi-Newton methods in general) is more difficult convergence and the requirement for a larger number of iterations to achieve a desired tolerance compared to the full Newton method.

The complex step derivation formula is a second-order accurate approximation for every real valued function of real variable $f(x)$:

$$\partial f/\partial x = Im(f(x+ih))/h$$

and can produce very accurate results for arbitrarily small values of the complex step h mainly because the subtraction operation (thus the roundoff error) present in finite difference formulas is avoided.

The coupled ODE system is integrated using the variable order method for stiff differential equations that is implemented in the MATLAB® function *ode15s*.

3.2 An Optimization Scheme for Load Capacity and Stability

The objective function is chosen to be equal to the real part of the dominant eigenvalue pair. For solution of the eigenvalue problem, the MATLAB® function *eig* is used. For the construction of the objective function, the model configuration and an initial fixed point solution are given as constant parameters with the actuator displacements and/or stiffness and damping properties as inputs. Then, Broyden's method is used for the calculation of the actual fixed point which varies with the objective function inputs. Once the fixed point has been evaluated, the eigenvalues of the Jacobian matrix are evaluated and the objective function returns the desired output value.

The actuator displacements are constrained between 60% of the radial clearance in the inward direction and 60% in the outward direction in a global sense. However for any given particular optimization problem, a large design space for the actuator displacements can lead to the objective function being unable to compute the new fixed point from the initial given value. To overcome this difficulty, in every optimization problem, the actuator displacements are constrained between ±20% of the initial actuator displacement vector q_0 for which the fixed point x_0 is already known. Once the optimization problem has converged on a solution e.g. q_1 which corresponds to a fixed point x_1, it is repeated with the new initial conditions and new constraints $\pm 20\% q_1$. This process is repeated until a convergent solution that satisfies the "global" constraints ($\pm 60\% c_r$) has been evaluated and the rotating speed is incremented.

The nonlinear optimization problem is solved using the function *surrogateopt* from the *MATLAB Global Optimization Toolbox®* which is recommended for computationally expensive objective functions.

4 Results and Discussion

In this section, a number of results from the reference system in which all the actuator displacements are equal to zero, and a number of case studies for different number of actuators, and actuator stiffness and damping properties, are presented. Two systems are analyzed, the properties of which are presented in Table 1 along with all relevant physical properties.

4.1 Calculations of the System with the Reference Configuration

The derived TEAD model reference calculations of equilibria, and transient behaviour are presented in this section for the two Bearings D = 30 mm and D = 100 mm. A sensitivity analysis is performed for the selection of the mesh intervals and it was found that a mesh as coarse as $N_x \times N_z = 24 \times 8$ provides an adequate level of accuracy for this inherently qualitative study with an acceptable amount of computational effort.

For the two reference systems with unbalance, a transient analysis is performed, see Fig. 2. The two systems are stable until a certain speed (approximately 1800 rad/s for the D30 System and 1100 rad/s for the D100 System) where the gas whirl phenomenon occurs and the period-1 (synchronous) motion experiences a period-doubling bifurcation, and the motion becomes of period-2, retaining this quality up to a higher rotating speed (c.a. 4100 rad/s for the D30

Table 1. Reference Model properties and values for the two investigated systems

	Model Properties & Units	D30 System	D100 System
Bearing parameters	Journal diameter $D = 2R$ [mm]	30	100
	Journal mass m_j [kg]	5	0.5
	Maximum Rated Speed Ω_r [krad/s]	6.5	20
	Radial clearance c_r	$R/500$	
	Length to Diameter Ratio L/D	1	
	Foil thickness h_f [μm]	100	
	Foil material Young's modulus E [GPa]	210	
	Foil material Poisson's ratio ν	0.3	
	Foil material density ρ [kg/m^3]	7860	
	Heat transfer coefficient β [W/m^2]	100	
Gas properties	Specific Gas Constant R_s [J/(kg K)]	287.05	
	Specific Heat Capacity c_{v0} [J/kg]	718	
	Viscosity at $T_0 = 273K$ μ_0 [kg/(m s)]	$1.716 \cdot 10^{-5}$	
	Sutherland's Constant S_μ [K]	111	
	Ambient Pressure p_∞ [Pa]	101325	
	Ambient Temperature T_∞ [K]	293	

system, and 1900 rad/s for the D100 system). Increasing the speed further will render a chaotic response, and shortly after the stiff integrator fails to converge. The period-2 motion is verified by running a transient analysis at a constant speed and calculating the Poincaré recurrence map, and the power spectral density using the Fast Fourier Transform, see Fig. 3.

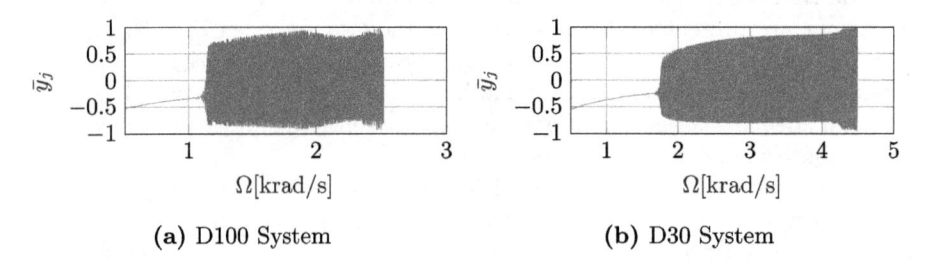

(a) D100 System

(b) D30 System

Fig. 2. Journal displacement in vertical direction during runup for (a) D30 System and (b) D100 System, with reference foil configurations, rotating acceleration $\alpha = 400$ rad/s^2 and unbalance grade $G2.5$.

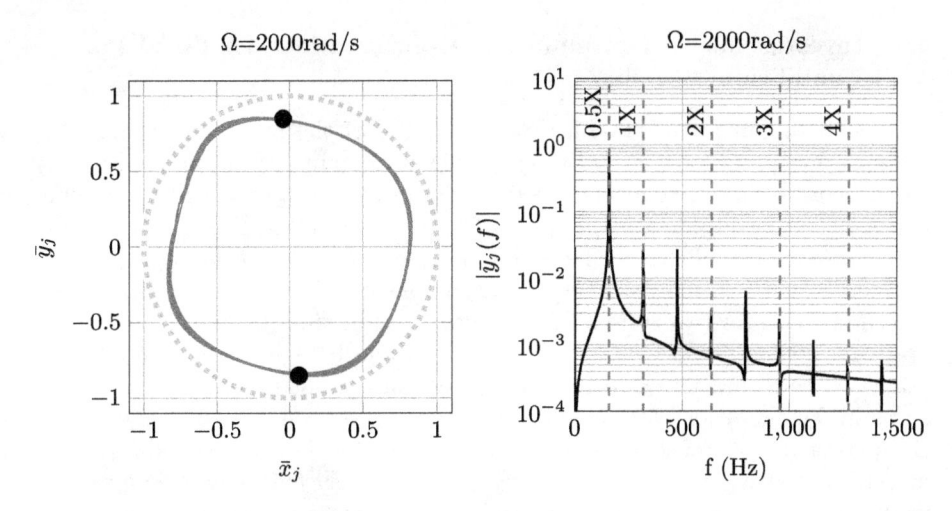

Fig. 3. Journal nondimensional orbit during a stable period doubling limit cycle, sychronous Poincaré recurrence map and corresponding power spectral density for D100 System.

The stability of the two systems with reference foil configuration is investigated for various bearing numbers $\Gamma = \mu R L^3/(m_j \sqrt{g c_r^5})$, calculated by varying the journal mass m_j, see Fig. 4. It is shown that lightly loaded AGFBs (larger bearing number Γ) will become unstable at lower speeds. To exhibit the potential for increasing the onset speed of instability, the most lightly loaded bearings (the ones corresponding to Table 1) are used in this study.

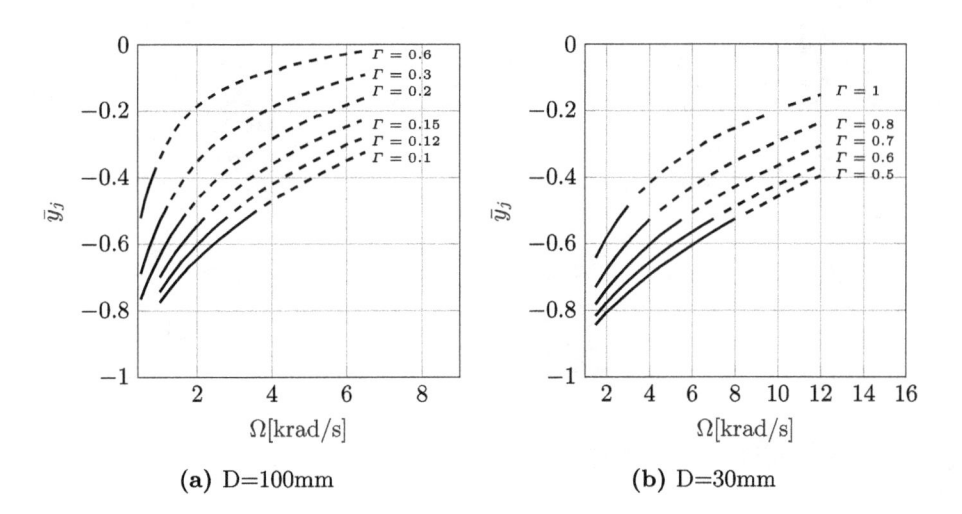

(a) D=100mm

(b) D=30mm

Fig. 4. Stability of equilibria for various bearing numbers Γ.

4.2 Investigation of Optimum Foil Configurations for Fixed Point Equilibrium Stability

In order to improve the stability characteristics of the reference systems, the optimization procedure, as described previously, is applied at various rotating speeds, corresponding to a range of DN values from 0.9e6 to 6.2e6 (A DN number of 6.55e6 corresponds to the speed of sound in air at 20 °C where the inertia effects become significant). The actuator positions, foil deformation, pressure and temperature profiles at the bearing midplane and the journal position are shown in Fig. 5 for two systems. It is shown that in lower speeds the evaluated optimum configuration can undergo change while between higher speeds, the evaluated configurations appear similar. Also important to note are the differences in configurations for similar DN numbers between the two systems and the qualitative differences between the configurations. The corresponding nondimensional density, pressure, film thickness foil displacement and temperature fields, and the deformed foil shape are shown in Fig. 6. The calculated actuator displacements for the D100 systems are applied with a ramp profile over the course of 1 s at three different rotating speeds, as shown in Fig. 7 and the subsychronous large-amplitude vibrations are suppressed both in the speed where the optimization was performed as well as in higher rotating speeds (provided the fixed point eigenvalues remain stable). The effect of the subsychronous large amplitude vibration suppression on the maximum film temperature is shown in Fig. 8. The optimization procedure is repeated at the maximum rated speed for the two systems.

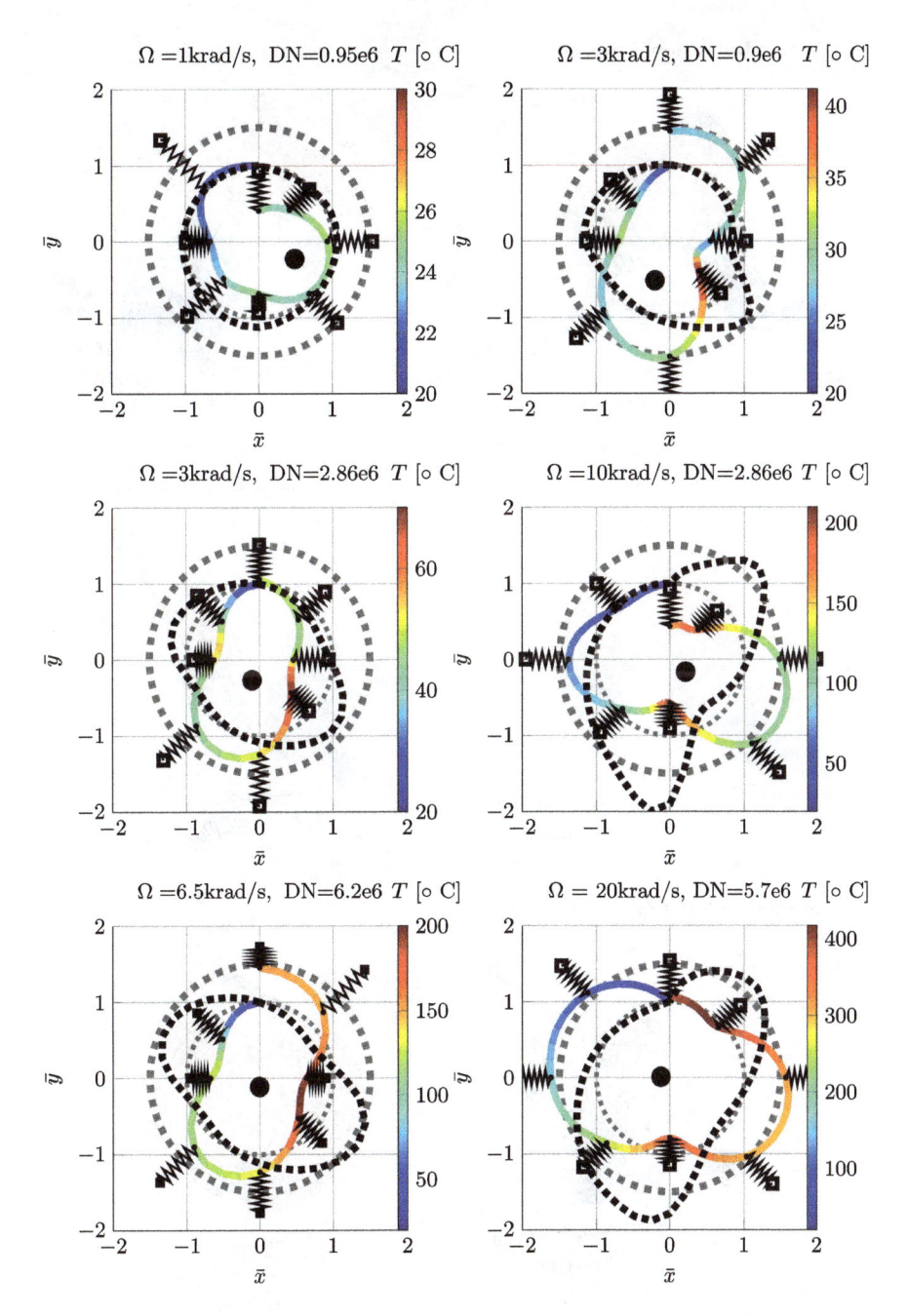

Fig. 5. Bearing midplane section with an applied optimum foil configuration with temperature and pressure distribution (black dotted line) along the circumferential direction for D100 System (left column) and D30 System (right column).

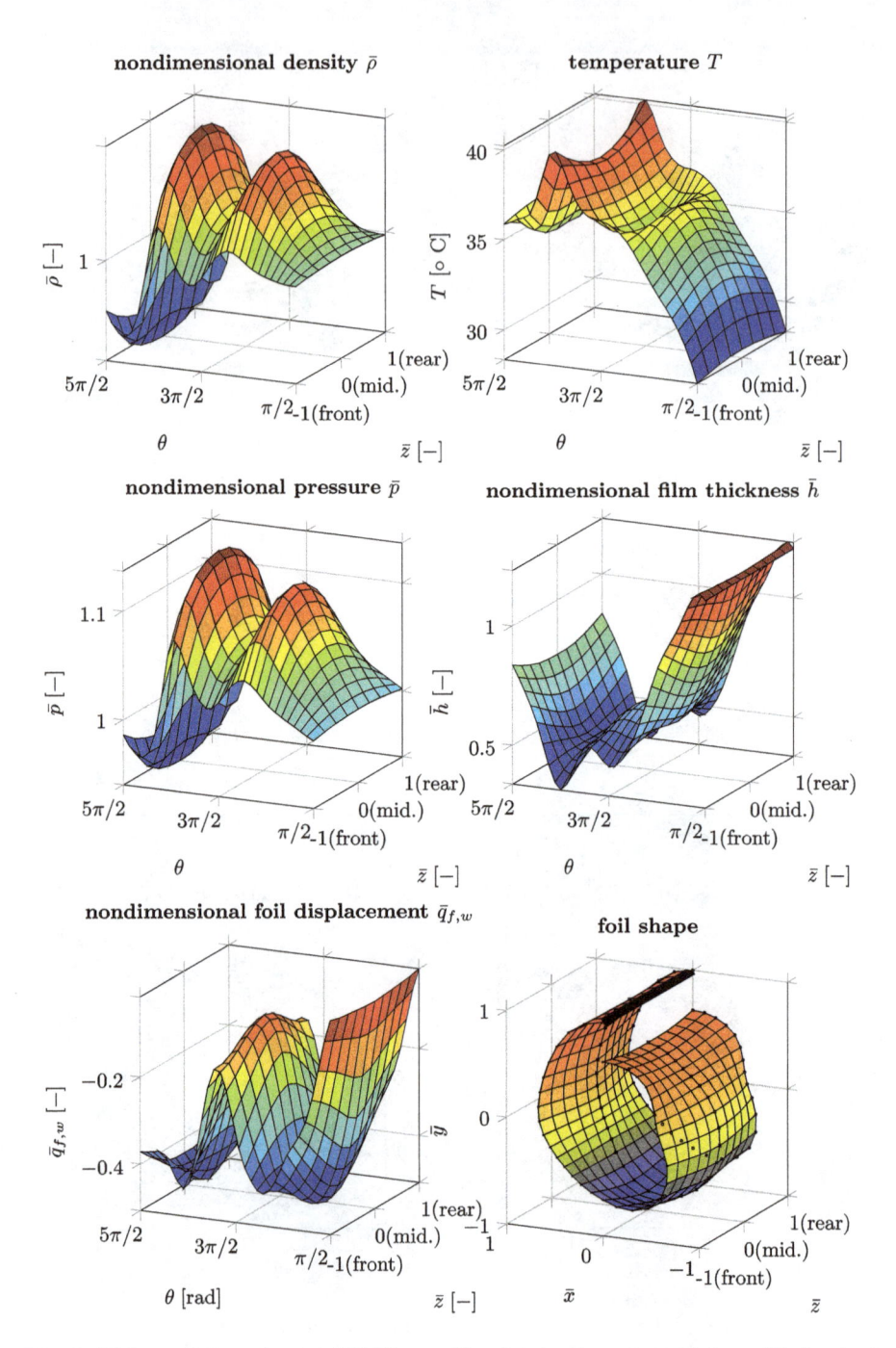

Fig. 6. Values of the relevant TEAD gas film lubrication states (at equilibrium) at $\Omega = 1500$ rad/s for an optimized foil configuration.

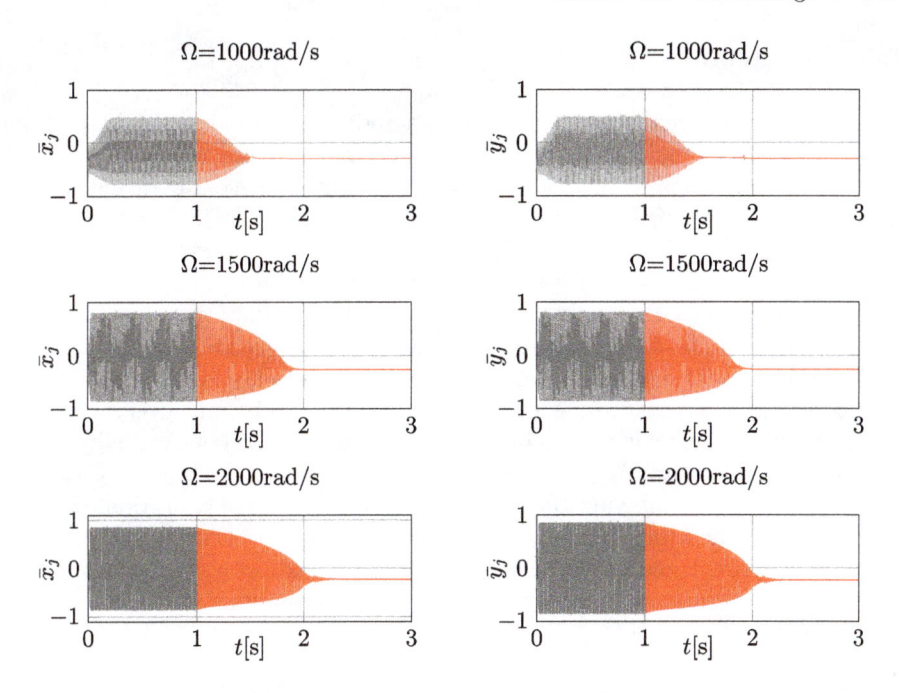

Fig. 7. Horizontal and vertical nondimensional journal displacement during suppression of subsychronous vibrations (red) at three different speeds by applying optimum configuration for D100 System.

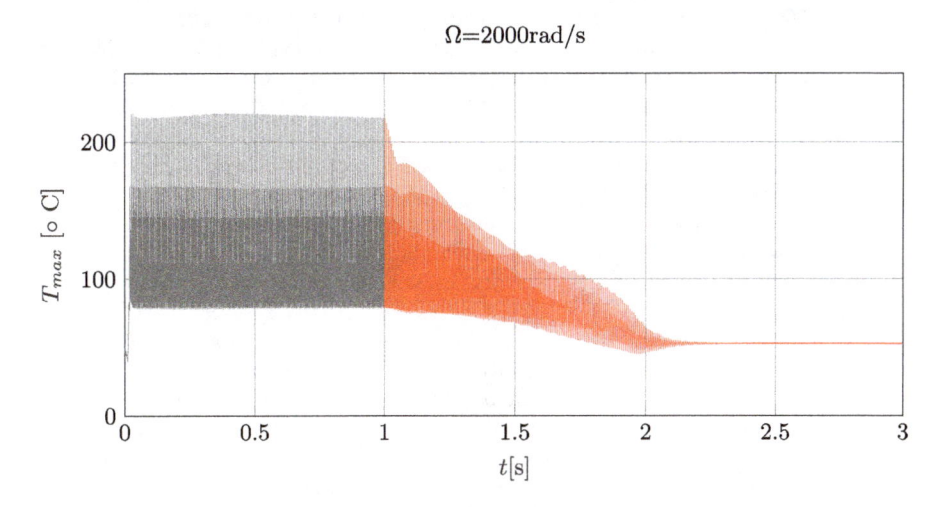

Fig. 8. Time series of the maximum film temperature during subsychronous vibration suppression for D100 System.

5 Conclusions

In this work, a predictiveThermo-elasto-aerodynamic model (TEAD) of an Active Gas Foil Bearing was used to investigate optimum foil configurations for increasing the onset speed of instability in a wide range of operating speeds. Two systems with different properties where investigated in similar DN values. It was found that it is possible to significantly increase the onset speed of instability in both systems by actively configuring the foil shape, in this study using simplified models of PZT actuators acting directly on the top foil, while completely omitting the bump foil and therefore all damping induced through the dry friction contact between top and bump foil as is the case in conventional Gas Foil Bearings. The potential of stabilizing large amplitude limit cycles occuring in the nonlinear model in speeds higher than the onset speed of instability (OIS) is also investigated and it was found that it is possible to stabilize self excited orbits by gradually applying the evaluated optimum foil configurations. Future work includes the implementation of more complex rotor models, the development of an optimal control law based on the present findings and an investigation of the optimum configurations for varying bearing properties.

References

1. Kim, T., San Andrés, L.: Effects of a mechanical preload on the dynamic force response of gas foil bearings: measurements and model predictions. Tribol. Trans. **52**(4), 569–580 (2009). https://doi.org/10.1080/10402000902825721
2. Sim, K., Lee, Y.B., Kim, T.H.: Effects of mechanical preload and bearing clearance on rotordynamic performance of lobed gas foil bearings for oil-free turbochargers. Tribol. Trans. **56**(2), 224–235 (2013). https://doi.org/10.1080/10402004.2012.737502
3. Schiffmann, J., Spakovszky, Z.S.: Foil bearing design guidelines for improved stability. J. Tribol. **135**(1), 011103 (2012). https://doi.org/10.1115/1.4007759
4. Walter, F., Sinapius, M.: Influence of aerodynamic preloads and clearance on the dynamic performance and stability characteristic of the bump-type foil air bearing. Machines **9**(8) (2021). https://doi.org/10.3390/machines9080178, https://www.mdpi.com/2075-1702/9/8/178
5. Sadri, H., Schlums, H., Sinapius, M.: Investigation of structural conformity in a three-pad adaptive air foil bearing with regard to active control of radial clearance. J. Tribol. **141**(8), 081701 (2019). https://doi.org/10.1115/1.4043780
6. Feng, K., Guan, H.Q., Zhao, Z.L., Liu, T.Y.: Active bump-type foil bearing with controllable mechanical preloads. Tribol. Inte. **120**, 187–202 (2018). https://doi.org/10.1016/j.triboint.2017.12.029, https://www.sciencedirect.com/science/article/pii/S0301679X1730587X
7. Guan, H.Q., Feng, K., Cao, Y.L., Huang, M., Wu, Y.H., Guo, Z.Y.: Experimental and theoretical investigation of rotordynamic characteristics of a rigid rotor supported by an active bump-type foil bearing. J. Sound Vibr. **466**, 115,049 (2020). https://doi.org/10.1016/j.jsv.2019.115049, https://www.sciencedirect.com/science/article/pii/S0022460X19306121

8. Park, J., Kim, D., Sim, K.: Rotordynamic analysis of piezoelectric gas foil bearings with a mechanical preload control based on structural parameter identifications. Appl. Sci. **11**(5) (2021). https://doi.org/10.3390/app11052330, https://www.mdpi.com/2076-3417/11/5/2330
9. Sadri, H., Schlums, H., Sinapius, M.: Design characteristics of an aerodynamic foil bearing with adaptable bore clearance. In: Proceedings of the ASME Turbo Expo: Turbomachinery Technical Conference and Exposition, vol. 7B: Structures and Dynamics (2018). https://doi.org/10.1115/GT2018-76204, https://doi.org/10.1115/GT2018-76204. V07BT34A031
10. Park, J., Sim, K.: A feasibility study of controllable gas foil bearings with piezoelectric materials via rotordynamic model predictions. J. Eng. Gas Turbines Power **141**(2), 021027 (2018). https://doi.org/10.1115/1.4041384
11. ISO-21940:16(E): Mechanical vibration—Rotor balancing—Part 12: Procedures and tolerances for rotors with flexible behaviour. Standard, International Organization for Standardization, Geneva, CH (2016)
12. Leister, T.: Dynamics of rotors on refrigerant-lubricated gas foil bearings. Ph.D. thesis, Karlsruher Institut für Technologie (KIT) (2021). https://doi.org/10.5445/IR/1000130548
13. White, F.: Viscous Fluid Flow. McGraw-Hill International Edition. McGraw-Hill, New York (2006)
14. Moran, M.J., Shapiro, H.N., Boettner, D.D., Bailey, M.: Fundamentals of Engineering Thermodynamics, 8th edn. Wiley, Hoboken (2014)

Gyroid Lattice Structures for Tilting Pad Journal Bearings

Ludovico Dassi(✉)[iD], Steven Chatterton[iD], Paolo Parenti[iD], Andrea Vania[iD], and Paolo Pennacchi[iD]

Department of Mechanical Engineering, Politecnico di Milano, Via G. La Masa 1, 20156 Milan, Italy
{ludovico.dassi,steven.chatterton,paolo.parenti,andrea.vania, paolo.pennacchi}@polimi.it

Abstract. Hydrodynamic journal bearings are standard supporting elements for industrial rotating machines. Radial load is sustained by the pressure field generated in the oil film wedge between relative moving surfaces, due to hydrodynamic lubrication phenomenon, but shear stress heats up the lubricant, limiting the maximum allowable load. Continuously growing of specific power increases the compactness and the efficiency of machines, with positive effects on the environmental footprint of production plants. This approach requires an innovative design procedure of the bearing components. The aim of this paper is to design and produce a new solution of pad for Tilting Pad Journal Bearings with a cooling circuit shaped inside the component to reduce the oil film temperature. The heat transfer has been relevantly increased using gyroid lattice from bioinspired inspiration. A comparative analysis is performed by changing the relevant parameters for lattice geometry and a complete thermo-mechanical study is performed with numerical and experimental analysis. Such a complicated geometry has been manufactured by polymer-metal feedstock extrusion, an innovative Metal 3D printing technology. Printability limits, local defects and macro geometrical errors are investigated and quantified to make the design procedure reliable. Finally, this research activity opens new opportunities: gyroid structure is smart solution for light weighting of structural parts and the employment of copper could increase significantly heat exchange efficiency.

Keywords: Journal bearings · gyroid lattice · Bound Metal Deposition

1 Introduction

Fluid film bearings are standard solution for supporting large shaft lines of industrial machineries. This technology relies on hydrodynamic lubrication principle which guarantees low friction and higher stability with respect to other solutions. The working principle is based on the convergent shape between two relatively moving surfaces, in which the resulting pressure distribution sustains the load [1].

The behavior of the lubricant film is strongly influenced by the local temperature field. Frictional heat is generated by shear stress in the oil film and this reduces lubricant

---

© The Author(s), under exclusive license to Springer Nature Switzerland AG 2023
A. Chasalevris and C. Proppe (Eds.): ABROM 2022, LNME, pp. 150–161, 2023.
https://doi.org/10.1007/978-3-031-32394-2_10

density and viscosity with detrimental effects on bearing performances, as discussed in [2]. Moreover, coating materials, like Babbitt metal, generally have a limited melting point, so the temperature field has to be carefully controlled to avoid damages.

Generally, cooling is performed by injection of cold oil at pad leading edge, as discussed by Hagemann et al. in [3] but a more efficient solution consists in the internal cooling of pads, as discussed by Najar et al. in [4].

The aim of this paper is to develop an improved layout of internal circuit, taking the advantages of pioneering gyroid lattice [5]. After a bibliographical investigation, recently discovered biological structures are considered for the application since gyroid lattice features high mechanical resistance and increased heat exchange properties. These are the key aspects to design a compact heat sink for a structural component, like the pad. In this sense, Kaur et al. summarize effectively the characteristics and the advantages of 3D printed heat exchangers [6].

Eventually, once all information is gathered and analyzed, pad design can be addressed. The proposed cooling circuit is characterized by a suitable path that covers uniformly the sliding surface; the channel is internally filled with lattice geometry to increase the surface/volume ratio and so enhancing the heat convection term. Comparing the numerical simulations and experimental tests, differences between nominal and real geometries are captured and discussed. Moreover, the innovative solution is compared to the baseline cooled pad, featured by a 6 squared multi-channels cross section, already developed by the same research group and discussed in [7].

Gyroid lattice is a very complex 3D shape and standard machining techniques are not able to produce it. Nevertheless, gyroid has the important property of self-sustainability and it can be easily printed without supports, as discussed in [8]. Metal 3D printing technology is a convenient solution for manufacturing of the pad and, in particular, polymer-metal feedstock extrusion simplifies the task. Basically, the production is performed in three steps. The first one is the printing of the polymeric and wax filament (binder) bonded with metal powder. Then, chemical debinding is performed in a solvent bath and finally thermal debinding and sintering is done in a furnace. This setup allows to speed up production rates by processing parts in small benches and it avoids local thermal effects on materials.

2 Methods

2.1 Gyroid Structure

Gyroid lattice is a geometry inspired from nature, studied the first time by Alan Schoen in 1970s. The structure is noticed in C.Rubi butterfly's wings and it guarantees lightness and good mechanical performances [9]. From the mathematical point of view, gyroid belongs to the Triple Period Minimal Surface (TPMS), family of curves which minimize the surface energy for a given boundary. As a result, lattice is characterized by smoothness with no edges or corners. The mathematical expression is reported in Eq. 1.

$$\varphi(x, y, z) = cos(x)sin(y) + cos(y)sin(z) + cos(z)sin(x) = 0 \quad (1)$$

There are two possible methods to generate the gyroid structure starting from the minimal surface, as shown in Fig. 1. The former solution is based on surface thickening and it creates a *sheet based* TPMS structures. Two parallel surfaces are generated by offset from the nominal one, creating a solid surface wall with a homogeneous thickness. The latter is based on the solidification of one of the two volumes defined by the minimal surface, so that creating a *skeletal based* TPMS structure.

Relevant parameters to fully define gyroid shape are: lattice type (sheet or skeletal), unit cell length (edge of the minimal constitutive cell), relative density (percentage of solid volume in an elementary cube) and external lattice dimensions (obtained by the repetition of constitutive cell in a periodic structure).

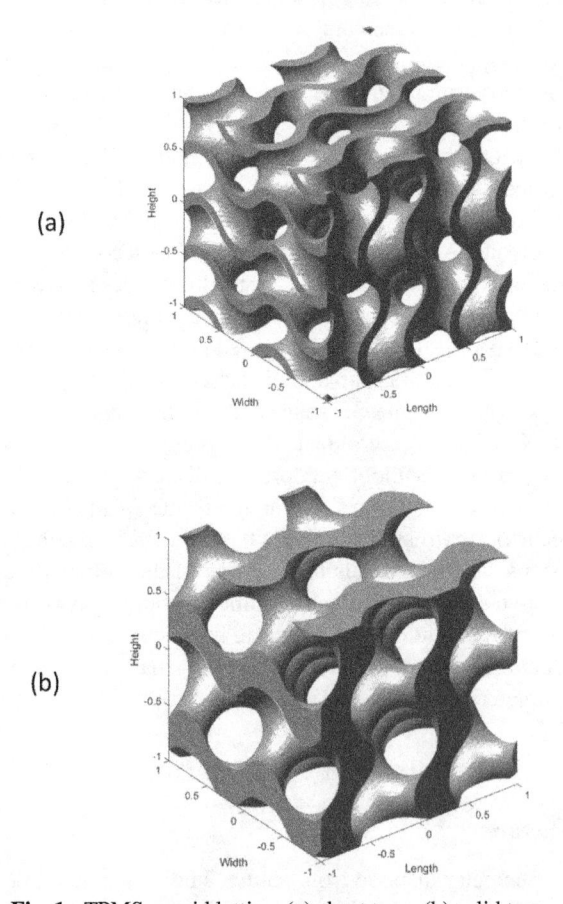

Fig. 1. TPMS gyroid lattice: (a) sheet type, (b) solid type.

Since literature is quite lacking in information, the first step of this research is to study the thermo-mechanical properties of Gyroid lattice, namely the compressive and elastic modulus for static structural analysis and relevant fluid dynamics properties, as discussed in [10]. Pad geometry bears low load condition (around *3MPa* distributed pressure field), so the mechanical properties are not so limiting factors.

Numerical simulations are performed on pad geometry, ranging different kind on lattices, to assess the resulting compressive stiffness. The relative compression strength and relative Young modulus are almost independent from the considered material. They are referred to mechanical properties of bulk material and they are sensible to relative density. Broadly speaking, gyroid lattice assures 5% of relative Young modulus and 10% of relative Compressive Yield Strength, as discussed by Al-Ketan in [11].

2.2 Fluid Dynamic Analysis

First of all, the macroscopic fluid dynamics properties of gyroid structure are investigated, by mean of numerical and experimental analysis. Some cylindrical samples are exploited to simplify the problem and a complete thermo-fluid dynamics study is performed to obtain the best counterbalance on performances (pressure drop and heat exchange) for pad application, by considering the same setup proposed in [12]. Tubes are 50 mm long (1/4 of pad cooling circuit length), while the free cross-sectional area to the fluid is kept equivalent to the baseline pad geometry (to make easy comparisons). A parametrical study is conduced to compare the lattice geometries and to quantify the effect of parameters settings (cell length, relative density, working fluid, flow rate and solid material).

Unit cell size is fixed to 8 mm for printability limits, both solid and sheet types are considered, relative density is ranged between 15–45%, ISOVG68 oil and water are considered as working fluid, while for the solid part steel and copper have been investigated.

In the model, coolant flow rate is ranged and the pressure drop on the lattice is estimated. The temperature at fluid inlet (40 °C) and external cylindrical wall (60 °C) are imposed, so that the integral heat flux is calculated to match different lattices in the same conditions. Some representative quantity fields of the samples are computed, as shown in Fig. 2.

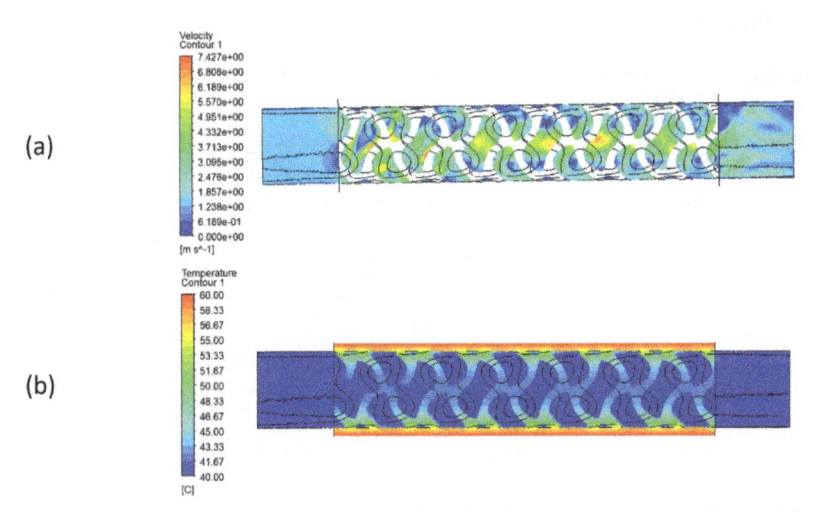

Fig. 2. Fluid-dynamics analysis of sheet 35% cylindrical samples: (a) velocity field, (b) temperature field.

Finally, numerical simulations on pressure drop are compared with experimental campaign tests, once samples are printed, to validate the model and to capture differences between ideal geometries and the real ones.

The experimental setup consists in a flushing facility to measure pressure drop on real lattice samples, as shown in Fig. 3. The circuit is composed by a control unit with the oil pump and a temperature control loop to keep lubricant at 40 °C. Temperature and pressure probes are used for measures while flow rate is checked with a digital scale.

Fig. 3. Flushing experimental set-up.

Measures are repeated five times for robustness and statistical analysis is performed to define the second order interpolating polynomials of pressure curve.

2.3 Pad Design

The lattice type more suitable for the application is found to be the sheet one with relative density 35% because it guarantees the highest heat exchange with respect to its pressure drop. The proposed cooling circuit is characterized by a suitable path that covers uniformly the sliding surface (M-shape serpentine), the internal domain of the cooling channel is filled with gyroid lattice in order to increase the surface/volume ratio enhancing the heat convection term. Pad structure is made of stainless steel SS316L with relevant mechanical performances and corrosion resistance, while the sliding surface is coated with standard white metal SnSb8Cu4, a common choice for Tilting Pad Journal Bearings.

A CFD numerical model is implemented to predict thermal and fluid dynamic performances of the proposed solution. Figure 4 shows the cross sections of velocity and temperature field for the simulation.

(a)

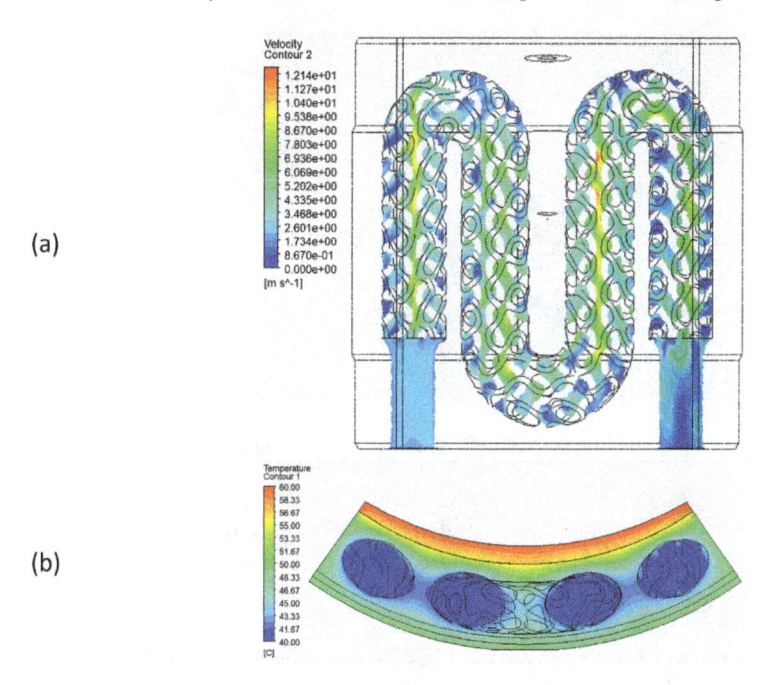

(b)

Fig. 4. Cross sections of sheet 35% lattice pad

2.4 3D Printing of Pad

To produce such a complicated shape, traditional manufacturing processes are not helpful but Metal 3D printing Additive Manufacturing has reached the required affordability to create functional prototypes, ready for industrial applications. Namely polymer-metal feedstock extrusion, a.k.a. Bound Metal Deposition technology (BDM), does not need of powder bed printing, avoiding the difficult task to flush away the residual powder from the internal channels. Moreover, it allows the direct printing of copper, which is a promising further development for pad cooling. A manufacturability study is carried out on stainless steel SS316L to define the printing limits and important aspects to take care during printing. Cylindrical samples also allow the printability of lattice to be verified. They are trials to setup the best printing profile on the machine to avoid defects.

Samples geometries are firstly considered, virtual modes are processed, printing preferences are defined and slicing of geometries is set. Six samples are produced firstly printing them and then debinding and sintering them in two benches. Once printed, accurate visual inspections search for evident defects, porosity is checked via Archimedean weighting and nominal dimensions are measured. An example of the printed samples is shown in Fig. 5.

Pad production is more difficult, overall geometry is defined but some corrections are needed to obtain a successful pad manufacturing. During the design phase common rules for AM design have been already considered, like printing tolerances, overhang angles, no need of teardrop holes, part stability, aspect ratio and ceramic stringing. Then,

Fig. 5. Cylindrical tube filled with gyroid lattice for fluid-dynamics characterization.

excess metal of 1 mm is added over all surfaces for finishing operations. Finally, some venting holes are created to allow a correct debinding of the part.

The first pad prototype is printed vertically. This solution has a good surface finishing quality but it is affected by many cracks parallel to layer planes. The cracks are caused by internal stresses of printing phase and they can be avoided by a slightly tilting of slicing plane of the pad (around 20°). In this way principal stress directions are no more aligned with layer planes, avoiding delamination phenomenon. An example of the printed pad is shown in Fig. 6.

Fig. 6. 3D printed pad with internal cooling circuit (as printed state).

Printed pads are finished by standard milling operations on the external surfaces, coolant holes are threaded and white metal coating is applied on the sliding surface. Dimensional and geometrical quotes are checked and no significant distortions are found. Finally the pad is ready for the application and it is used for flushing experiments. The mechanically finished pad is shown in Fig. 7.

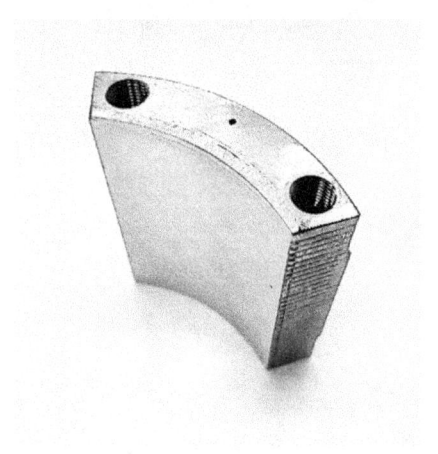

Fig. 7. 3D printed pad after mechanical finishing operations.

3 Results

Generally speaking, gyroid lattice is confirmed to be an effective solution to create compact heat exchangers. The numerical results show great opportunities but production process variability has to be taken in consideration.

For pad application, main goal is to significantly increase the heat exchange while counter pressure is a soft constrain and it is chosen lower than 5 bars for technical simplicity. In the case of oil as cooling fluid, sheet lattice have better exchange than solid ones, sheet 45% has excessively high pressure drops while sheet 35% at 4L/min seems to be a good trade-off for the application (Q: *+41%*, Δp: *+4bar*) with respect to baseline, 6 channels with oil). Sheet 35% lattice is chosen and SS316L as material. All the comparisons are made keeping the same hydraulic power as reference condition. Figure 8 directly compares the pressure drop and heat exchange of sheet 35% pad and baseline solution as function of the flow rate.

Then, performance gain can be boosted furthermore by changing materials. Sheet 35% layout still is a good choice (6L/min flow rate) and it guarantees a profitable tradeoff of heat exchange and pressure drop with respect to baseline by using water as cooling fluid (Q: *+50%*, Δp: *+2.7bar*). The performances can be again increased by using water and copper (Q: *+1636%*, Δp: *+2.7bar*) Actually, lattice works like a finned geometry and conduction efficiency becomes a key aspect of the problem. Experimental pressure drop curve is reported in Fig. 9 for oil; measured values show great repeatability and perfectly fit Bernoulli 2-order polynomial.

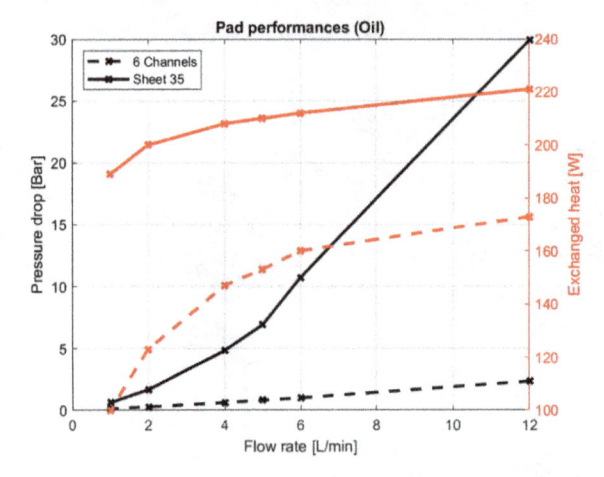

Fig. 8. Numerical simulation on pressure and heat exchange properties with oil.

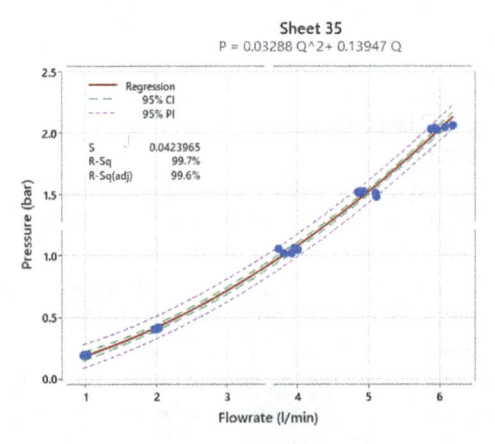

Fig. 9. Experimental test on sheet 35% circular sample with oil.

Comparing CFD and experimental measures, an averaged difference of 10% is present, as shown in Fig. 10. Further investigations demonstrate that it is caused by printing errors in the relative density.

Concerning the static mechanical analysis, pad geometry is studied under nominal loading conditions, which are described in detail in [13]. The bell shape pressure distribution is taken from a previous work (Fig. 11) and it is applied on the top sliding surface. Pad is clamped on the bottom pivot surface and the average temperature of the system (50 °C), is imposed constant over the pad to simplify the problem. No hydrostatic pressure is considered inside the cooling channel surfaces. By looking at the deformation results, the proposed solution guarantees the same stiffness of baseline layout and no stress intensification is visible inside lattice domain. The deformation field computed by numerical FEM analysis is shown in Fig. 12. The maximum deformation in micrometers of pad structure, under nominal load, are listed in Fig. 12 (Table 1).

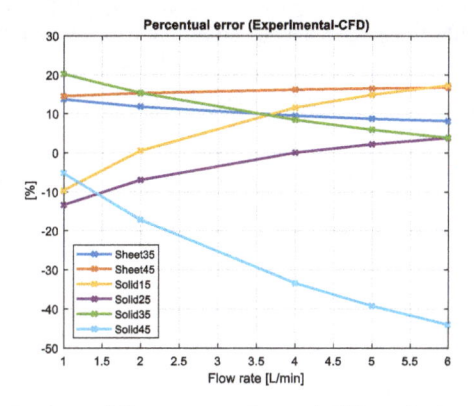

Fig. 10. Pressure difference experimental-CFD on lattice samples.

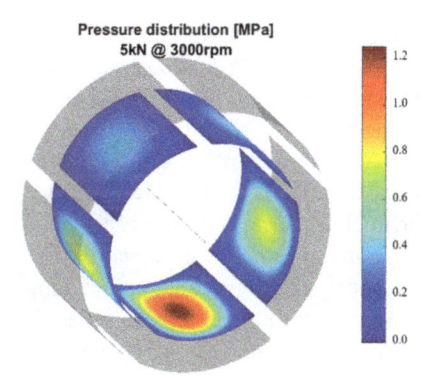

Fig. 11. Bearing assembly pressure distribution.

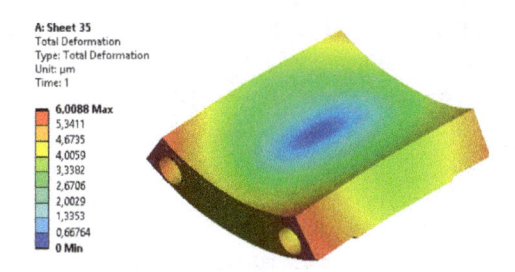

Fig. 12. Pad deformation field.

Table 1. Comparison on vertical compliances of pad.

Internal layout	Max deformation [μm]	[%]
6 Channels	6.77	-
Sheet 35%	6.00	−12.8%
Sheet 45%	4.11	−71.8%
Solid 15%	7.70	+13.7%
Solid 25%	6.85	+1.2%
Solid 35%	6.21	−3.5%
Solid 45%	4.31	−68.5%

4 Conclusions

The proposed research activity proposes a successful re-design of pad for hydrodynamic bearings. The exploitation of recently discovered gyroid lattice allows to better cool down the structure and increase the loading performances. Static mechanical response and fluid dynamics properties have been addressed in detail throughout both numerical and experimental analysis. Pad application has large convenience in this sense, innovative internal channels can decrease the sliding surface temperature with favorable effects on oil dynamic viscosity, film thickness and rotating speed. This work favors the worldwide trend of increasing the performances and the efficiency of machines in the engineering world. Great effort is also spent studying the manufacturability of pad. Bound Metal Deposition is able to create lattice complex geometries and structural sections of the component. A defects-free part is obtained balancing advantages and drawbacks of this AM technology and adapting the production set up consequently. A great advantage stands in the absence of excess powder to be flushed away, which could be a difficult task for a long cooling channel.

Further developments are highlighted during design development phase. Firstly, it will be interesting to validate the pad performances on a rotating test rig. Then, use of copper is confirmed to be an effective solution to particularly enhance properties of gyroid lattice and finally the comparisons between CFD and real pressure measures are a good indicator to estimate the effective printing accuracy.

References

1. Stephen, W., Giddings, C., Lee, S.: Characterization of hydrodynamic lift forces by field-flow fractionation. Inertial and near-wall lift forces. Chem. Eng. Commun. **130**(1), 143–166 (1994)
2. Fillon, M.: Thermal and deformation effects on tilting-pad thrust and journal bearing performance. Presentato al Congresso Ibérico de Tribologia (2005)
3. Hagemann, T., Zeh, C., Schwarze, H.: Heat convection coefficients of a tilting-pad journal bearing with directed lubrication. Tribol. Int. **136**, 114–126 (2019). https://doi.org/10.1016/j.triboint.2019.03.035
4. Najar, F., Harmain, G.: Performance characteristics in hydrodynamic water cooled thrust bearings. J. Tribol. **10**, 28–47 (2016)

5. Michielsen, K., Stavenga, D.G.: Gyroid cuticular structures in butterfly wing scales: biological photonic crystals. J. R. Soc. Interface **5**(18), 85–94 (2008). https://doi.org/10.1098/rsif.2007.1065
6. Kaur, I., Singh, P.: State-of-the-art in heat exchanger additive manufacturing. Int. J. Heat Mass Transf. **178**, 121600 (2021)
7. Chatterton, S., Pennacchi, P., Vania, A., Dang, P.V.: Cooled pads for tilting-pad journal bearings. Lubricants **7**(10), 92 (2019). https://doi.org/10.3390/lubricants7100092
8. Parenti, P., Puccio, D., Colosimo, B.M., Semeraro, Q.: A new solution for assessing the printability of 17-4 PH Gyroids produced via extrusion-based metal AM. J. Manuf. Process. **74**, 557–572 (2022). https://doi.org/10.1016/j.jmapro.2021.12.043
9. Khaderi, S.N., Deshpande, V.S., Fleck, N.A.: The stiffness and strength of the Gyroid lattice. Int. J. Solids Struct. **51**(23), 3866–3877 (2014)
10. Luo, J.-W., Chen, L., Min, T.: Macroscopic transport properties of Gyroid structures based on pore-scale studies: Permeability, diffusivity and thermal conductivity. Int. J. Heat Mass Transf. **146**, 118837 (2020)
11. Al-Ketan, O., Rezgui, R., Rowshan, R.: Microarchitected stretching-dominated mechanical metamaterials with minimal surface topologies. Adv. Eng. Mater **20**(9), 1800029 (2018)
12. Li, W., Yu, G., Yu, Z.: Bioinspired heat exchangers based on triply periodic minimal surfaces for supercritical CO_2 cycles. Appl. Therm. Eng. **179**, 115686 (2020)
13. Chatterton, S., Pennacchi, P., Dang, P.V., Vania, A.: A test rig for evaluating tilting-pad journal bearing characteristics. In: Pennacchi, P. (ed.) Proceedings of the 9th IFToMM International Conference on Rotor Dynamics. MMS, vol. 21, pp. 921–930. Springer, Cham (2015). https://doi.org/10.1007/978-3-319-06590-8_75

Behavior of an Active Magnetic Bearing as a Stern Tube Bearing: A First Approach via Simulations

Vasileios-Menelaos Koufopanos and Pantelis G. Nikolakopoulos[✉]

Department of Mechanical Engineering and Aeronautics, University of Patras, 26504 Patras, Greece
koufopanosv@upnet.gr, pnikolakop@upatras.gr

Abstract. Currently Journal Bearings are used at a vessel's stern tube. This means that the friction losses are significant, not to mention the material wear because of the contact, which is often a cause of frequent maintenance. The Active Magnetic Bearing's (AMB) technology offers a contact-free force generation through the magnetic field of electromagnets located around the carried shaft. Thus, friction losses or wear are avoided, while controlling the position of the shaft through a feedback control system in real time is enabled through coil current control. This paper presents a Finite Element Model (FEM) created in ANSYS Multiphysics software, which simulates the magnetostatic field of the AMB and calculates the bearing's load capacity. MATLAB Software is then used to simulate the feedback control system of the AMB, where a PID controller is chosen as the most appropriate, offering both stability and zero steady state error. Finally, the dynamics of the rotor – AMB System is simulated using a FEM model, with gyroscopic loads considered, and the system's critical speeds are calculated. A case study is executed for a low-weight high-speed propeller shaft of a catamaran. The results of this first approach through simulations show that the rotor – AMB system operates far below its first critical speed, thus an AMB can offer the requested load capacity for a low-weight propeller shaft, and that combined with a PID controller, the rotor can be carried and controlled within a normal range of functioning parameters' values.

Keywords: Active Magnetic Bearing · FEM · Propeller shaft · Rotor dynamics · PID controller

Nomenclature

$A =$ Cross-sectional area
$A_a =$ Air cross-sectional area
$A_{Fe} =$ Iron cross-sectional area
$A_m =$ Magnetic potential vector
$A_z =$ Magnetic potential vector in the Z-direction
$B =$ Magnetic flux density/induction
$D =$ Diameter

$d =$ Damping coefficient
$E =$ Young's modulus of elasticity
$F_{mag} =$ Magnetic force
$H =$ Magnetic field's magnitude
$I =$ Electric current/Inertia moment of a surface (2nd order)
$I_b =$ Bias current
$I_c =$ Control current
$k =$ Stiffness coefficient
$K_D =$ Controller's derivative gain
$K_I =$ Controller's integral gain
$k_i =$ Current stiffness coefficient
$K_P =$ Controller's proportional gain
$k_s =$ Displacement stiffness coefficient
$m =$ Mass
$N =$ Number of coil's turns
$P_{el} =$ Electric power
$q =$ Displacement vector
$s =$ Magnetic gap
$x =$ Displacement on the X-direction
$y =$ Displacement on the Y-direction
$\rho =$ Density
$\Omega =$ Rotational speed

1 Introduction

Active Magnetic Bearings (AMBs) take advantage of a magnetic force to support the load of a rotating shaft, without any mechanical contact.

When it comes to Stern Tube Bearings oil or sea water are used as lubricants to achieve Hydrodynamic lubrication and therefore a sliding between the bearing surface and the one of the rotating propeller shaft. The Hydrodynamic Bearing geometry and manufacturing are quite simple, while their load capacity is significant even when they have relatively small radial clearance. The stiffness and damping coefficients are high, thanks to the viscosity the lubricants have. As a result, Hydrodynamic Bearings are suitable even for heavy rotors, like the propeller shafts, and respond well in shock load cases. However, high load capacities also result to high friction losses and material wear. The frequent need for lubricant replacement has economic and environmental consequences, not to mention the pollution caused by a possible leakage of the oil. When sea water is used as a lubricant, the bearings operate mostly under boundary lubrication conditions as hydrodynamic lubrications is difficult to be achieved due to the low viscosity of water. Sverko and Sestan [1] studied the behaviour of an aft stern tube bearing and concluded that analytical methods, even for hydrodynamic lubrication gave significantly different results compared to the experiments performed, due to the shaft's deformation at that location.

The support of propeller shafts seems to have significant challenges to face. The development though, of magnetic levitation technology and mechatronic systems has offered an innovative solution to this problem. On the one hand there is the contactless force application, while on the other there is the involvement of the electric current in the mechanism, which is responsible for the magnetic field creation. This involvement, of an electric quantity in a mechanical system, is that turns the Hydrodynamic Journal Bearing, a passive machine element, to an Active Magnetic Bearing (AMB), a mechatronic system.

Overall, the AMBs offer a frictionless rotation, with no material wear and need for lubricant. This means, lower operational and maintenance costs, less environmental pollution, and longer lifetime for the mechanism's components. The absence of lubricant also allows higher rotational speeds while the AMB is more appropriate for harsher operational environments (even vacuum). The electric current supplied to the electromagnets can be controlled, thus vibration damping, accurate positioning and in general process monitoring are possible. However, we should not forget that the contact-free magnetic force requires power supply, while the necessary equipment for the system's control and monitoring will surely have a significant cost. Friction losses may be eliminated, but the electric and magnetic phenomena have their own ways of losing energy, such as magnetic hysteresis or eddy currents.

Perhaps the most complete source for the theory and the design principles of the AMBs is the book by Schweizer and Maslen [2]. Knowledge on this technology, though, is constantly expanding. According to Shelke and Chalam [3], who experimented on the optimization of the electromagnet's weight and energy loss reduction, the 8-poles layout (i.e. 2 pairs of electromagnets) was found to be the optimum. In scientific literature, however, one can find several other cases for the number of the poles. Zimon et al. [4] present a 12-pole non-symmetric AMB, accompanied with simulations and experiments for several rotational speeds of the carried shaft. Stimac et al. [5] describe a modelling method for an elastic rotor – AMB system, emphasizing on vibrations' elimination at the 1st critical speed. Pewekar et al. [6] created a FEM model and compared the results of its simulations with experimental measurements, aiming to design a journal AMB with optimum power consumption and efficient hardware. Farmakopoulos et al. [7, 8]) designed a Hybrid Journal Bearing, which can operate either as a Hydrodynamic Bearing, or as an Active Magnetic Bearing, or even both, depending on the circumstances. They initially present the simulation models created using the ANSYS and MATLAB/Simulink software, followed by CATIA drawings and some laboratory experiments. Bompos et al. [9] had presented the potential role of an AMB in ship propulsion.

The two most commonly used magnetic bearing types are the passive and the active magnetic bearings. Passive magnetic bearings show low power consumption and achieve the passive suspension of the rotor through the interaction between stator's and rotor's permanent magnets or the reluctance suction between the two [10–12]. However, this type of bearings lacks active control and has low damping and stiffness properties. Beside this, active magnetic bearings, creating force in terms of electromagnets, a feedback control loop etc. show special behavior in terms of high stiffness, good controllability and damping properties compared to passive bearings [13–16].

Stiffness is generally known to be one of the most important parameters for the whole AMB control system [17, 18] and it is categorized mainly into displacement stiffness and current stiffness. The investigators of [19] preset that both coefficients of current and displacement stiffness for a power magnetic bearing can be precisely determined through an application of an open-loop static levitation force on the rotor.

In the recent works [20–22] a proportional-integral-derivative (PID) controller and electromagnetic actuator was incorporated in a coupling misaligned flexible rotor-bearing model and the nature of vibration of the system was studied using both the numerical and experimental investigations.

An integration scheme between active magnetic bearings and journal bearings presented in [23]. A novel control scheme has been adopted to adjust the active magnetic bearing equilibrium position to coincide with the journal bearing equilibrium position, eliminating some problems in the operations such as the opposing forces acting on the journal/rotor from both bearings at static equilibrium.

The aim of this paper is to examine the behavior of the active magnetic bearings as stern tube bearings using a simulation procedure. To this direction, a first approach of the physics and feedback control of the rotor – AMB system are presented; using CAE models the physics and control of the AMBs are analyzed so that the required values of the main functioning parameters are specified. The aforementioned models are then applied to two AMB cases, both for static and dynamic system behavior; one that will act as the aft stern tube bearing for a low-speed propeller shaft of a tanker and one for a high-speed shaft of a catamaran. Contactless rotation is highly beneficial in this case because of the friction and wear problems stern tube bearings have. Additionally, the re-alignment of the shaft that can be achieved, will improve its operation at its entire length. However, proper sealing of the bearing is required, combined with corrosion protection.

The rest of the paper is organized as follows; Sect. 2 contains the mathematical description of the problem, and the main equations are explained. Section 3 describes the case on which the AMB will be studied; an AMB for a high-speed catamaran shaft is examined. Section 4 presents the CAE models (FEM and MATLAB) created specifically for this analysis, verified via examples from the literature. The models are then used for the simulations of the propeller shaft problem. Results are summarized and explained in Sects. 4.1, 4.2 and 4.3 while Sect. 5 contains the conclusions of this paper.

2 Active Magnetic Bearings: Theory and Basic Equations

2.1 Magnetic Circuits

Magnetic bearings are designed to generate contact-free forces by controlling the dynamics of electromagnets. They provide reaction forces for the weight of the rotor and can also control its position and vibration. In addition, they must give a satisfactory dynamic performance. The magnetic bearings used in this work have opposing coil pairs located at $\pm 45°$ from the vertical axis z. Consequently, the vectorial sum of the forces developed by the coils gives the total force generated by the magnetic bearings. Since this is an opposing pair configuration, one of the magnets depends on the control current I_c added to the bias current I_b (which means $I_b + I_c$), while the opposing one depends on the control current subtracted from the bias current $(I_b - I_c)$ (Fig. 1).

According to Schweizer and Maslen [2], a single axis force of the magnetic pair can be calculated as (Eq. 1):

$$F_{mag} = \mu_0 NA \left[\left(\frac{I_b + I_c}{s-z}\right)^2 - \left(\frac{I_b - I_c}{s+z}\right)^2 \right] \cos\theta \quad (1)$$

where μ_0 is the relative permeability of free space, A the pole face area, N the number of the coil windings, θ the polar angle of the coil, z the shaft displacement in the vertical direction and s the air gap when $z = 0$.

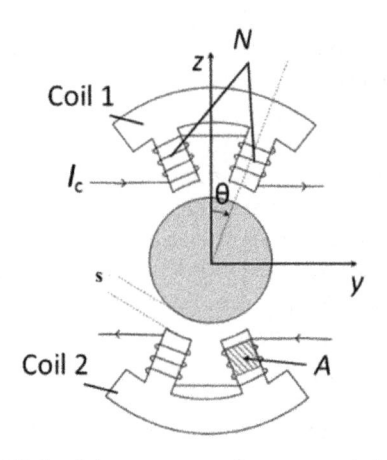

Fig. 1. Pair of electromagnets for a magnetic bearing

Linearizing Eq. 1 around $z = 0$ and $I_c = 0$ leads to (Eq. 2):

$$F_{mag} = k_i I_c - k_s x \quad (2)$$

where k_i is the current stiffness (Eq. 3):

$$k_i = \frac{\mu_0 N^2 A I_b \cos a}{s^2} \quad (3)$$

and k_s the displacement stiffness of the bearing (Eq. 4):

$$k_s = -\frac{\mu_0 N^2 A I_b^2 \cos a}{s^3} \quad (4)$$

2.2 Feedback Control System

Since, according to literature, a Rotor – AMB system is unstable [2], an appropriate feedback system needs to be added so that the system gets stabilized (Fig. 2). This feedback will consist of one or more sensors that will gather information regarding the position of the rotor in real time, and a controller which will supply the coils with the necessary control current for the adjustment of the magnetic force, based on the data coming from the sensors.

In order to define the values of the control system parameters (controller's gains) we follow the method provided by Schweizer and Maslen [2]. Technically, we need a positive stiffness factor that will act against the negative stiffness of the electromagnets, in addition to a damping factor that an electromagnet cannot offer.

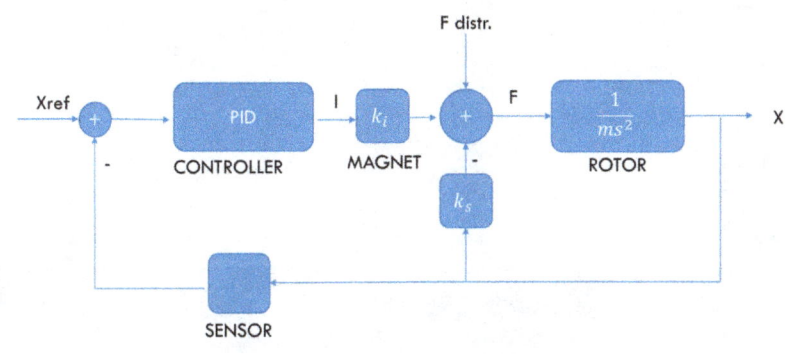

Fig. 2. Closed-loop Rotor - AMB system

According to literature [2] a PID controller is appropriate whose gains, the proportional K_P and the derivative K_D, are described by Eq. 5:

$$K_P = \frac{k - k_s}{k_i} \quad \text{and} \quad K_D = \frac{d}{k_i} \tag{5}$$

where k is the system's stiffness factor and d the damping factor. According to Schweizer and Maslen [2], a stiffness value between 1 to 3 times $|k_s|$ is suggested. As for the damping factor it is suggested not to exceed the product $2\sqrt{(mk)}$. The value of the integral gain K_I can be defined through a trial-and-error process, keeping always in mind the attributes of the controllers, already known from the feedback control systems' theory. The main system requirements are stability and zero steady state error (i.e. rotor placed exactly on the stator's center).

2.3 Dynamics of the Rotating Shaft

Since the shaft that is about to be examined (Fig. 3) combines large diameter with relatively short length ($L/D < 10$) the Timoshenko theory is the most appropriate for the rotor's dynamic analysis. This theory considers that the shaft's cross section does not remain vertical to the longitudinal axis, in addition to the impact of shear and torsional stresses.

For a shaft of an homogenous and isotropic material the equation which describes the relation between the vertical displacement y of a cross section as a function of its position x and time t, and the external force $q(x,t)$ is (Eq. 6):

$$EI\frac{\partial^4 y(x,t)}{\partial x^4} + \rho\frac{\partial^2 y(x,t)}{\partial^2 t} - \left(\gamma + \frac{EI\rho}{\kappa AG}\right)\frac{\partial^4 y(x,t)}{\partial x^2 \partial t^2} + \frac{\rho\gamma}{\kappa AG}\frac{\partial^4 y(x,t)}{\partial t^4}$$
$$= q(x,t) + \frac{\gamma}{\kappa AG}\frac{\partial^2 q(x,t)}{\partial t^2} - \frac{EI}{\kappa AG}\frac{\partial^2 q(x,t)}{\partial x^2} \tag{6}$$

where E is the Young's modulus of elasticity, I the cross-section's moment of inertia, ρ the linear density of the material, κ the shear stress factor and

$$\gamma = \rho \times \frac{I}{A} \tag{7}$$

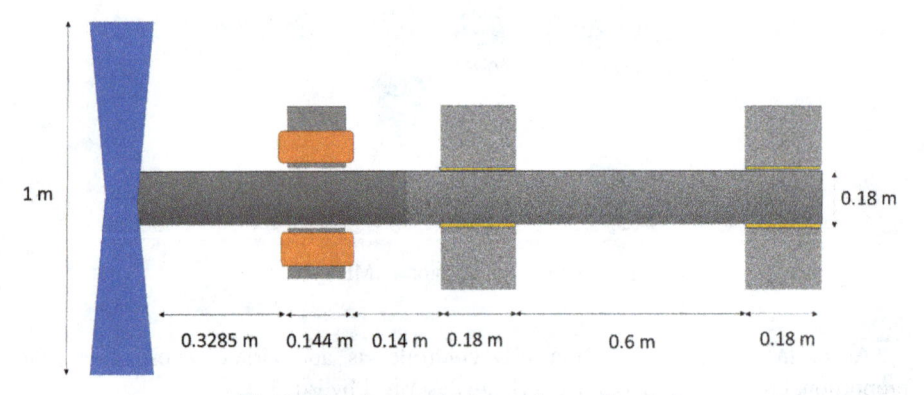

Fig. 3. Sketch of the high-speed propeller shaft

3 Case Study: An AMB for a High-Speed Propeller Shaft

In order to check the sensitivity of a propeller's shaft – AMB system's behavior, a relatively low weight ship rotor with high rotational speed was selected for the following analysis. The propeller shaft of a high-speed passenger craft, which rotates at 1940 rpm is considered (Fig. 3). The diameter of the shaft is 180 mm, the load which the AMB will need to carry is 2.5 kN and the length between the aft stern tube (AMB) and the propeller is 328.5 mm. The material of the bearing and the shaft is steel, with Young's modulus at 211 GPa, Poisson ratio at 0.3 and density 7800 kg/m³. The *B-H* curve can be found in [24]. For the rotor dynamics analysis propeller is modelled as a disk with 1 m diameter and 34.3 mm thickness.

4 Numerical Modelling

In this section, the simulation of the AMBs for the case described above is presented. The AMB will consist of 4 pairs of electromagnets, meaning 8 poles in total, the dimensions of which will depend on the shaft's geometry. As a material for the stator stainless steel is chosen considering that also the rotors' material is the same. For the coils' cables a 1.5 mm^2 cross section is considered as well as a 1.7×10^{-8} Ωm resistivity for the material. The part of the shaft's load that the AMB will need to support can be derived from the static analysis of the shaft. For the dynamic case, the shaft will be modelled as a cantilever beam carrying a disk at the point exactly where the propeller is located. The disk's mass will represent the whole mass that the AMB will need to carry, while its diameter will be equal to the propeller's.

Magnetic Circuits and Rotor Dynamics are solved with the Finite Elements Method performed by the ANSYS Software, while the Feedback Control System is simulated by MATLAB scripts. The ANSYS model for the magnetic circuits will be used for the magnetic force calculation, while the electric power consumed by the coils will also be calculated. The reliability of Eq. 2 is tested by performing several displacements on the rotor and changes on the current and by checking their effect on the magnetic force. Thus the factors k_i and k_s are calculated and the limits where the linear approach for the magnetic force is reliable are calculated. Feedback control then follows in order to make sure that the overshoot of the system's response is acceptable for a series of static loads. In addition, the behavior of the control current is simulated. The controller's gains are calculated from the Eqs. 5 and their values will be put to the test. Finally, through the rotor dynamics model and the Campbell diagrams the critical rotational speeds will be found and the modes of the shaft are presented. Results and parameter values are manually transferred from the one model to the other.

4.1 Magnetic Circuits

The Magnetic Circuits of the AMB were modeled in ANSYS. The problem of a magnetostatic field is solved using the Maxwell Equations and the Finite Elements Method. These equations form a different approach on the electromagnetic field than the one described above, as they derive from the principles of energy conservation and vector analysis. In particular, this problem needs the 2nd and the 3rd equation of Maxwell, which refer to the magnetic field.

The fact that the algorithm uses this approach for the solution of the magnetic circuit problem, makes the results more valid as the non-linearities that are neglected above are now taken under consideration. Furthermore, the materials are defined by their *B-H* curves, and, as a result, the hysteresis energy losses are also considered. These curves can be found in several handbooks like the one by the American Society of Metals [24].

After inserting the main geometric variables of the AMB, the geometry of the model can be created on the ANSYS Multiphysics environment, by creating nodes, lines, arcs and surfaces. Figure 4 shows the final geometry created, with each color representing a different material. A ferromagnetic material for the stator and its poles has been considered, as well as for a cylinder around the rotating shaft (it is assumed that because of the rotation the flux will only run in the rotor for a limited depth). For the air surrounding the poles (magnetic gap included) a magnetic permeability equal to μ_0 has been defined, as well as for the interior cylinder of the shaft which does not take any magnetic flux. The coils are copper, but since they are not part of the magnetic circuit, they also have the same permeability (μ_0).

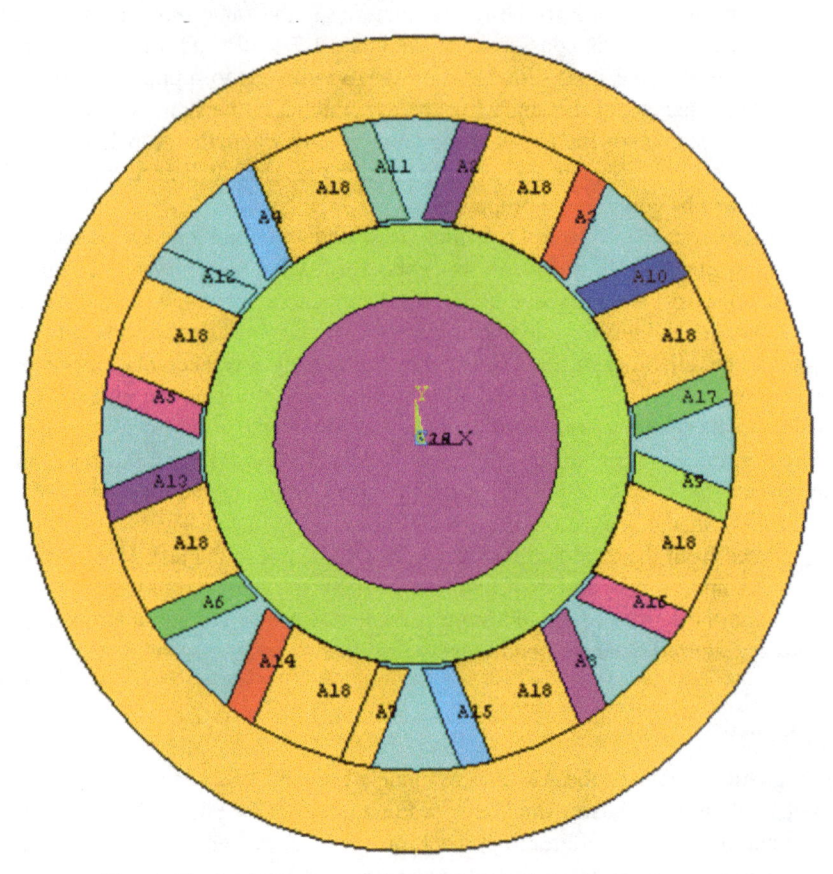

Fig. 4. Final AMB geometry in ANSYS Multiphysics Environment

Once the geometry has been created, meshing is the next step. For the elements size w the Automatic "Smart" Sizing option provided by the software is used, where the optimum setting ("Fine") is chosen. The mesh is formed by PLANE53 elements, which are the most appropriate for 2-D magnetic field simulations. They consist of 8 nodes and each node provides 4 Degrees of Freedom (DoF): z component of the magnetic vector potential (A_Z), time-integrated electric scalar potential (VOLT), electric current (CURR), and electromotive force (EMF).

The rotor is defined as a "component" which makes the results about the magnetic forces applied on it more accurate. The current supply is described by the current density value through the coils' cross section areas. The power supplied to the x-direction electromagnets is considered to be constant and equal to I_b, while the electromagnets of y-direction are supplied also with a constant control current in order to create a magnetic force to carry the weight of the shaft. The current for the upper electromagnet is $I_b + I_c$, and for the lower one $I_b - I_c$.

As a boundary condition, it is assumed that the flux runs exclusively in the stator's interior space (air and rotor included). Thus, the leakages outside the stator's diameter need to be zero, and that is why the magnitude of the magnetic vector potential A_Z is declared as zero for the nodes on the outer diameter of the stator.

Validation of the Model. Before using the model for the design problems, validation is needed. For that reason, a problem from a thesis by Antila et al. [25] is chosen and solved in order to compare the results. The material chosen for the stator and the rotor is Bochum V270, whose B-H curve points are given as inputs to the model. The input data of this problem is presented in Table 1.

Table 1. Input parameters for the magnetic circuits example

Parameter	Value
Stator's Diameter	164 mm
Bearing's Length	45 mm
Rotor's Diameter	93 mm
Magnetic Gap	0.44 mm
Coil Turns per pole	100
Pole Width	18 mm
Inner Stator's Diameter	127 mm
Thickness of rotor's ferromagnetic "ring"	14.35 mm
Bias current	2 A

The model runs for different values of the control current I_c (from 0 to 4 A) and for each solution it gives the results for the magnetic force applied on the rotor on the vertical direction. The value given in the results of the model is then compared to the value given by the linear equation of the magnetic force (Eq. 2) and the results presented by Antila et al. [25]. The comparison of those quantities is presented in Fig. 5.

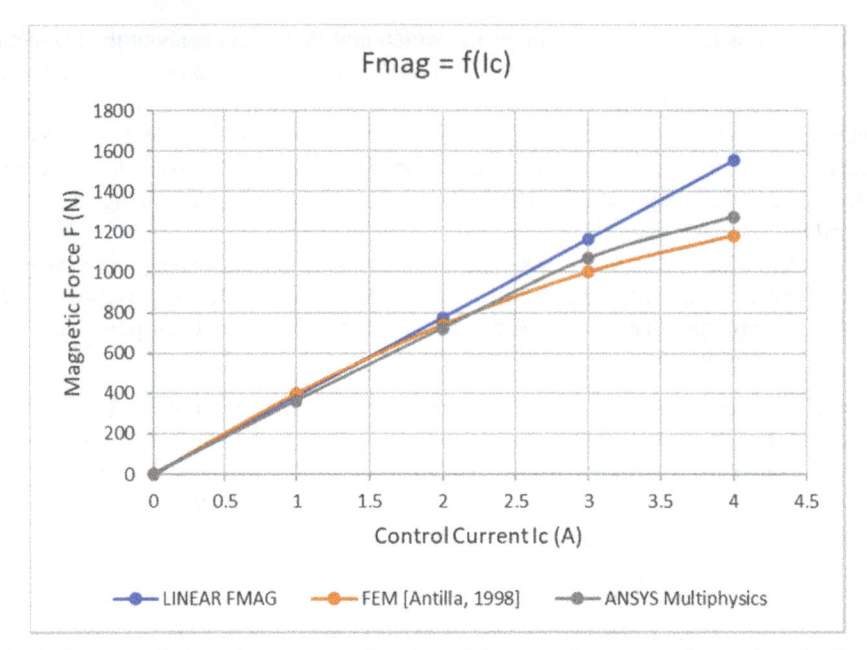

Fig. 5. Force applied on the rotor as a function of the control current, calculated by the linear equation, our FEM model and the model presented by Antila et al. [25].

From Fig. 5 it can be concluded that the model's results almost coincide with the ones of Antila et al. (p. 30 – Fig. 9b) [25]. It is also noticed that for control current values smaller or equal to 2 A (i.e. the value of I bias) the results almost coincide with the ones given by Eq. 2. This means that for small control current values the magnetic force can be calculated adequately by the linear equation.

Magnetic Circuits for the High-Speed Propeller Shaft Case. The main parameters of the AMB used for the high-speed propeller shaft case have the values presented in Table 2.

The vector graph of the magnetic flux density is presented in Fig. 6. Higher values can be noticed on the upper magnet where the current is $I_b + I_c = 1.25$ A, while lower ones on the lower magnet where the current is $I_b - I_c = 0.75$ A. On the weight direction (y-direction) the magnetic force value is calculated at 19865 N/m, so for 0.144 m of AMB length the force applied on the rotor is 2.86 kN.

This value is close enough to the weight of the rotor that the AMB needs to carry, so it is assumed that the parameters selected are appropriate. Further adjustment of the magnetic force is also possible through the feedback control system. Since the exact value of the current for each electromagnet is known, in addition to the cross section and resistivity of the coils' cables the electric power consumed for this case of magnetic levitation can be calculated. For this case and since the current values remain constant the power is calculated at 30.525 W.

Table 2. Input parameters for the magnetic circuits model of the high-speed rotor – AMB system

Parameter	Value
Stator's Diameter	500 mm
Bearing's Length	0.144 m
Rotor's Diameter	270 mm
Magnetic Gap	0.5 mm
Coil Turns per pole	800
Pole Width	60 mm
Inner Stator's Diameter	400 mm
Bias Current	1 A
(Constant) Control Current	0.25 A

This power value, however, neglects any ohmic resistance increase because of the temperature rise, or any magnetic power losses caused by the rotation of the shaft (i.e. hysteresis and eddy currents), as a result an experimental approach would definitely lead to a higher power consumption value.

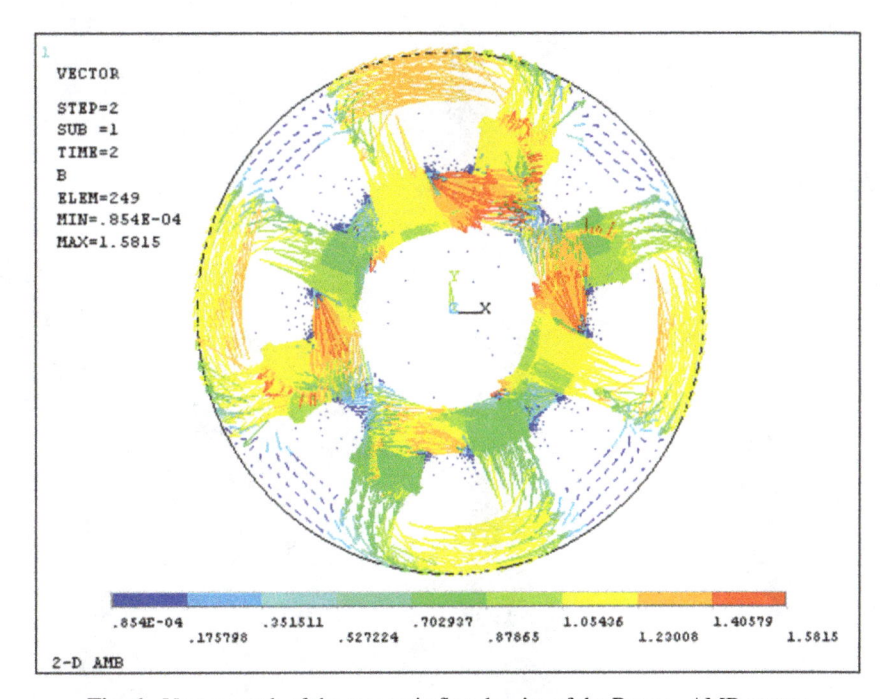

Fig. 6. Vector graph of the magnetic flux density of the Rotor – AMB system

Taking advantage of the non-linear solution the FEM can provide, the reliability of Eq. 2 can be checked. By applying several control current values from 0 to 0.8 A the k_i factor can be calculated and the maximum control current before saturation is reached can be determined. Results are shown as a graph on Fig. 7. For control currents smaller than 0.6 A the relation seems linear and the slope of the curve (i.e. k_i) can be estimated at 11165.28 N/A. From control current values above 0.6 A the slope decreases, and as a result power losses will occur.

In a similar way rotor displacements are applied (on y-direction) for the calculation of k_s. Results are shown as a graph on Fig. 8. Even for high displacement values (60% of the air gap) the force shows a linear behavior and the slope (i.e. k_s) can be estimated at 10322328.6 N/m. The decrease of the air gap is expected to cause an exponential increase in the magnetic force, something that does not seem to happen for the values considered in this case. However, it should be noticed that for 0.2 mm displacements and on the convergence of the FEM started decreasing, so the results may not be so accurate for higher values.

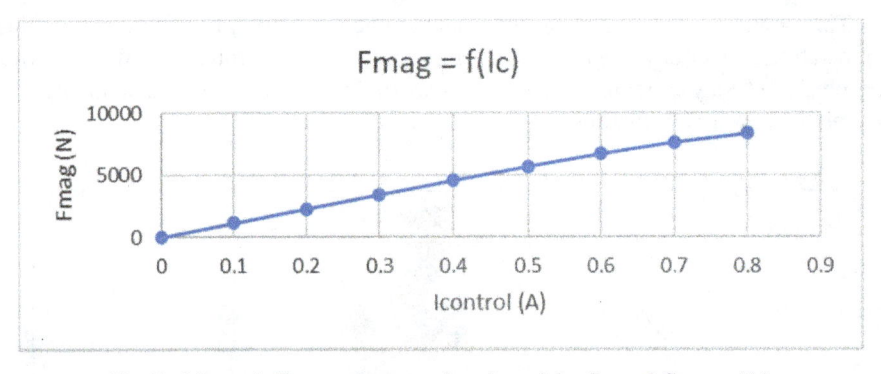

Fig. 7. Magnetic Force values as a function of the Control Current (Ic)

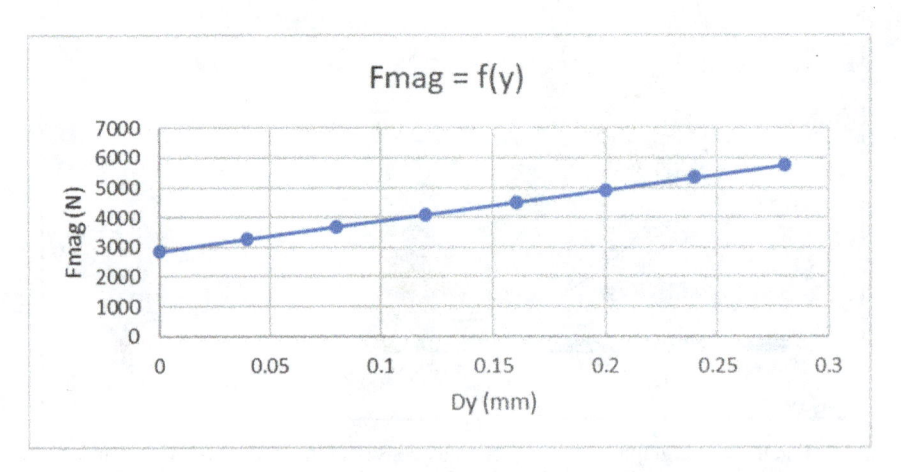

Fig. 8. Magnetic Force values as a function of rotor's displacement (dy)

4.2 Feedback Control System

The model of the feedback control system that completes the AMB structure is simulated in MATLAB. The static case is addressed, which means that rotor's length and rotational speed are not taken into account while the external loads applied are static (step loads). This is sometimes the case, when external loads displace the shaft as a whole body and have a constant value as long as they are applied (quasi-static loads). This is also the case of the shaft's weight, which is constantly applied on the body and needs a control current/ force to be supported. Additionally the length of the shaft from the AMB to the propeller is short compared to the diameter and while the rotational speed is far below the first critical, angular displacements could be neglected in a first approach.

Based on the theoretical approach already explained, the model that is going to be simulated will have a PID controller (Fig. 2). The model simulates the problem in a single direction, but as the structure is symmetric (control current for the weight support is set to be the reference/zero value) the same model can be used for each of the directions, vertical or horizontal.

The proportional and the differential gain of the controller (K_P and K_D respectively) can be determined using the method provided by Schweitzer and Maslen [2] explained in Sect. 2 (Eqs. 5). The integral gain (K_I) will be determined through trial and error considering that very high values could lead to instability of the system. The current and displacement stiffness factors (k_i and k_s respectively) have been calculated in the previous sections with FEM simulations. The limits within which the linear magnetic force approach is valid should be considered in every simulations. That means that overshoot values for the displacement higher than 0.2 mm and control currents above 0.6 A are not acceptable for the current case.

Validation of the Model. A case presented by Schweizer and Maslen [2] is chosen in order to check the validity of the model's results. The input data used is presented in Table 3.

Table 3. Input parameters for the feedback control system example

Parameter	Value
Rotor's mass	0.1 kg
Displacement Stiffness Factor	−10000 N/m
Current Stiffness Factor	10 A/m
Controller's Proportional Gain	5000 A/m
Controller's Integral Gain	800000 A/ms
Controller's Derivative Gain	6.32 As/m
Disturbance Force	100 N

The step response is shown in the graph of Fig. 9. As it became clear from the calculations in the previous section the system gets stabilized when a PID controller is used, while the steady state error is also eliminated. Additionally, the graph shows the same behavior with the example presented by Schweizer and Maslen (p. 46 Fig. 2.14) [2] so the results of the model can be considered reliable.

Feedback Control System for the High-Speed Propeller Shaft Case. The propeller shaft undertakes several loads from the water which are difficult to be modelled accurately. They are considered to be pseudo static and to have very large magnitude, while only the vertical and horizontal direction of their total vector is examined. A simulation is performed for a step load equal to the force calculated by the FEM model (2.86 kN) in order to check the behavior of the rotor for a load approximately equal to its weight and additionally to check the value of the control current, which should be equal to the one that was set to the FEM model.

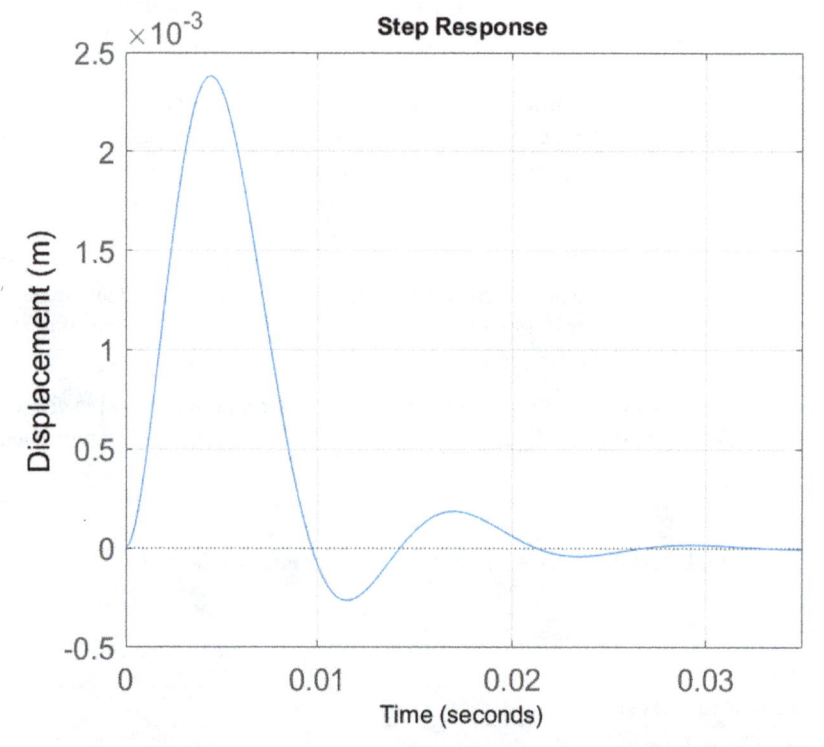

Fig. 9. System's step response for a disturbance force F = 100 N (PID Controller used)

A PID controller will be used, whose gains will depend on the damping and stiffness needed. For $k = -2k_s$ and $d = 2\sqrt{(mk)}$ the gains are calculated from Eqs. 5, except for the integral gain whose value is selected after a trial-and-error process (Table 4).

Thanks to the symmetry, simulation is needed only for a single direction. The system's response for a load equal to 2.86 kN, is presented in Fig. 10. The system is stable, and the disturbance is eliminated in 0.08 s. The overshoot (100 μm) is 5 times smaller than the air gap, within the acceptable range explained above. The control current's behavior for the same case is presented in Fig. 11. Its value in steady state is 250 mA, equal to the control current value used in the FEM model. This value is also within the acceptable range so that the magnetic saturation is avoided.

Table 4. PID Controller's Gains

Gain	Value
Proportional (K_P)	2774 A/m
Integral (K_I)	300000 A/ms
Derivative (K_D)	9.1 As/m

Several values of step load have also been simulated so that the maximum load can be determined, for which both the rotor's overshoot and the control current stay within the acceptable range. Results for the control current are presented graphically in Fig. 12 and for the maximum overshoot in Fig. 13. From the graphs it can be concluded that the rotor – AMB system can undertake load disturbances even as high as 5 kN, while for higher loads the overshoot reaches a non-acceptable value. This practically means that the rotor will get very close to one of the magnets.

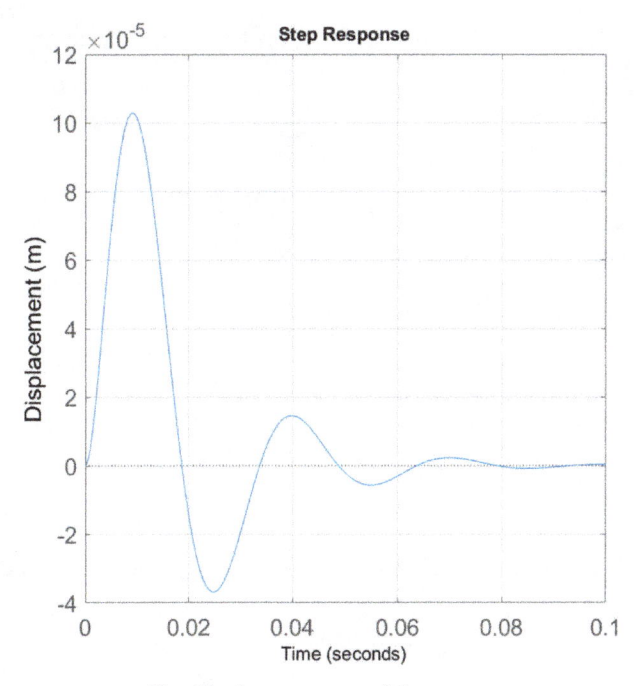

Fig. 10. Step response of the system

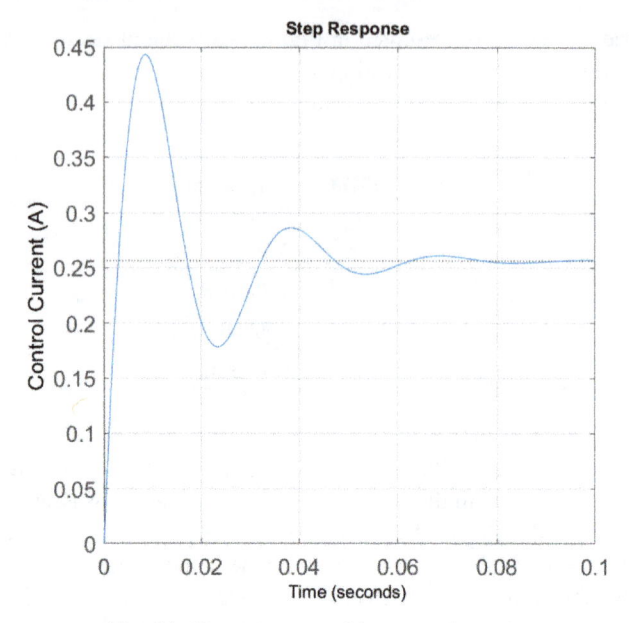

Fig. 11. Step response of the control current

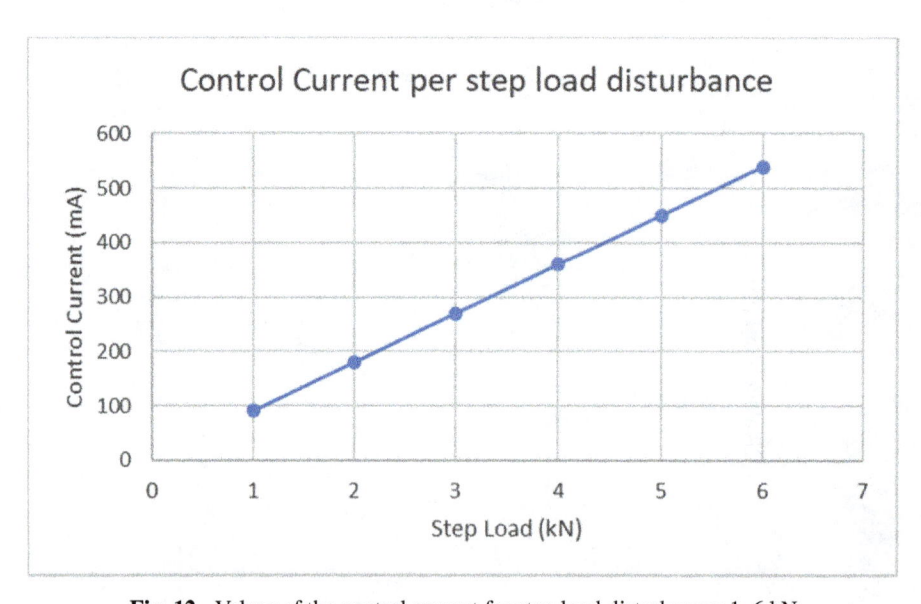

Fig. 12. Values of the control current for step load disturbances 1–6 kN

4.3 Dynamics of the Rotating Shaft

The dynamics of the Rotor – AMB System is simulated via the Mechanical APDL software (a.k.a. ANSYS Multiphysics). This FEM software provides elements which

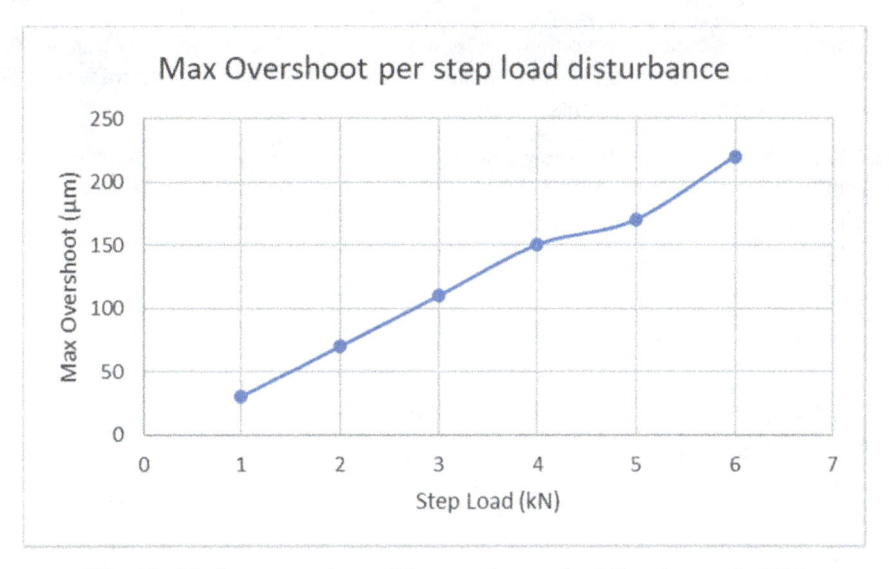

Fig. 13. Maximum overshoot of the rotor for step load disturbances 1–6 kN

can take bending and/or torsional deformations, elements with stiffness and damping factors and mass/inertia elements.

In particular the shaft is modelled with BEAM188 elements. Those are 3-D elements with 3 nodes and 6 DOFs in each node: 3 transversal and 3 rotational around the axes x, y and z of the local coordinates system. They enable torsional and bending vibrations analysis using the Timoshenko theory. The input parameters of the element are the radius of the shaft's cross section $D/2$, Young's Elasticity modulus E, Poisson's ratio PR, and the material's density ρ. The BEAM188 elements are defined between two nodes, the coordinates of which need to be defined.

The disk, which represents the mass of the rotor, is modelled with MASS21 elements. Those elements are dimensionless and are located on certain pre-existing nodes. They provide 6 DOFs, 3 transversal and 3 rotational displacements. They are defined by mass and moment of inertia values.

As for the bearings, they are modelled with COMBI214 elements. Those are 3-node elements which have as inputs stiffness and damping factors in matrix form. The elements of the matrices are the values of the factors in 8 different directions. Those values can be found in handbooks when it comes to hydrodynamic journal bearings, but for the AMB case, they can be calculated through current and displacement stiffness factors combined with the proportional and derivative gains of the controller:

$$k_{AMB} = k_i K_P + k_s \quad \text{and} \quad d_{AMB} = k_i K_D \tag{8}$$

These elements are located on nodes, which are purposely defined and geometrically coincide with previously defined nodes. More information about the element types mentioned above can be found at [26].

The boundary conditions confine the torsion of all the elements and any transversal motion along the rotation axis. Additionally, any motion, transversal or rotational, is confined for the bearing elements.

Validation of the Model. For the validation of the model, an example from a paper of Dumitru et al. [27] was used. It is the case of a rotating shaft supported by two roller bearings (Fig. 14). The shaft carries one disk. The input data of the problem is presented in Table 5.

Table 5. Input parameters for the Rotor Dynamics example

Parameter	Value
Young's modulus of elasticity	2×10^{15} N/m^2
Shaft's material density	1000 kg/m^3
Shaft's diameter	10 mm
Shaft's length	s + a = 0.3 + 0.4 m
Disk's mass	30 kg
Rotatory moments of inertia	1.2 kgm^2
Polar moment of inertia	1.8 kgm^2
Bearings' stiffness factors	k_{zz} = 0.7 N/m, k_{yy} = 0.5 N/m

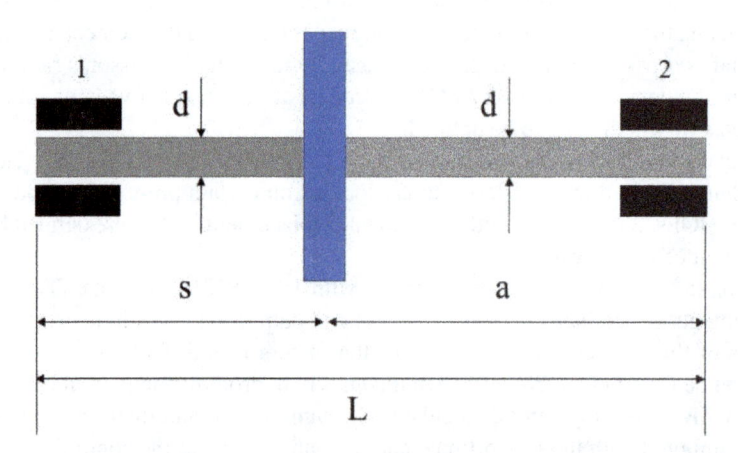

Fig. 14. Geometry of the rotor – bearings system

CAMPBELL DIAGRAM

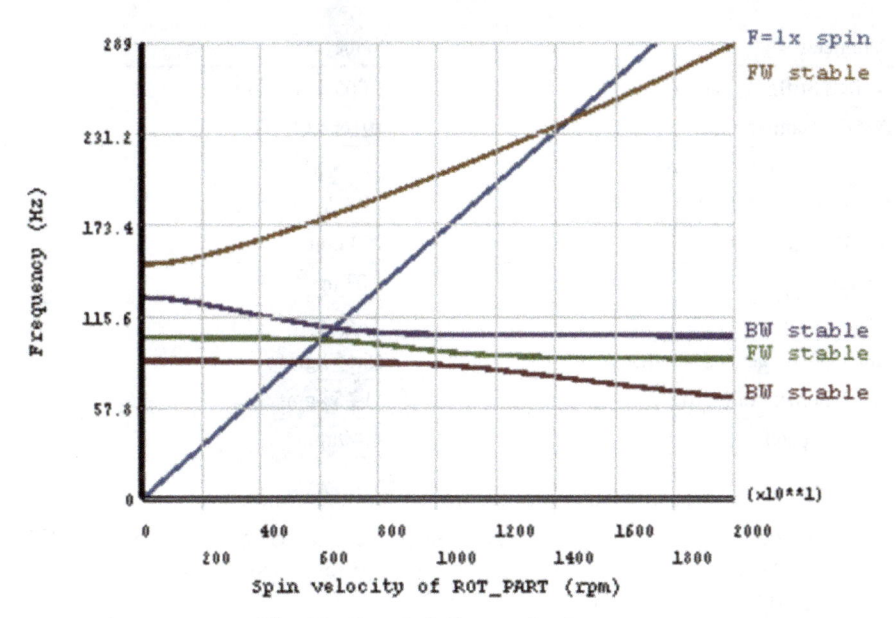

Fig. 15. Campbell diagram for the rotor

The results of the analysis give a Campbell diagram for the rotational speeds 0–20000 rpm (Fig. 15). The diagram of the current model is almost the same with the one given by the example [27]. The only difference is that the 4th eigenfrequency in this diagram, although it has a similar behavior, it takes lower values than the ones it takes in the example. Since both cases deal with numerically modelled and solved problems, some minor differences in the results can be considered acceptable and therefore the model is valid.

Dynamics of the High-Speed Propeller Shaft. For the Dynamics of the propeller shaft simulation the rotor presented in Fig. 3 was used. This rotor is supported by an AMB (as an aft stern tube bearing) and by two journal hydrodynamic bearings whose stiffness and damping coefficients were chosen for the Someya et al. handbook [28], for bearings that operate at a 0.131 Sommerfeld number and L/D ratio equal to 1. As for the AMB, its stiffness and damping factors are calculated from Eqs. 8. The input parameters of the model have the values presented in Table 6.

Table 6. Input parameters for the rotor dynamics model of the propeller shaft – bearings system

Parameter	Value
AMB's Stiffness Factor	2.065×10^7 N/m
AMB's Damping Factor	101604.05 Ns/m
Young's modulus of elasticity	211×10^9 N/m^2
Shaft's material density	7800 kg/m^3
Shaft's diameter	180 mm
Shaft's length	1.07 m
Disk's mass	250 kg
Rotatory moments of inertia	15.65 kgm^2
Polar moment of inertia	31.25 kgm^2
Rotor's speed	1940 rpm

The model created contains 54 BEAM188 elements, 1 MASS21 and 3 COMBI214 elements. The solution will be performed for a speed range from 0 to 5000 rpm (speed step 50 rpm) Gyroscopic loads are considered.

Table 7. Critical Speeds of the propeller shaft

Critical Speed number	Critical Speed Value (Hz)
1st	79.602
2nd	87.358
3rd	289.69
4th	383.55

The results presented in Table 7 (Critical Frequencies) and Fig. 16 (the Campbell diagram) show that the shaft operates far below the 1st critical speed (it operates at 1940 rpm) and as a result gyroscopic loads and vibrations because of propeller imbalances will not affect significantly the behavior of the rotor inside the AMB. This result indicates also that our feedback control approach focusing more on heavy static loads was more vital for an initial analysis. In Figs. 17, 18, 19 and 20 the modes of the propeller shaft are presented.

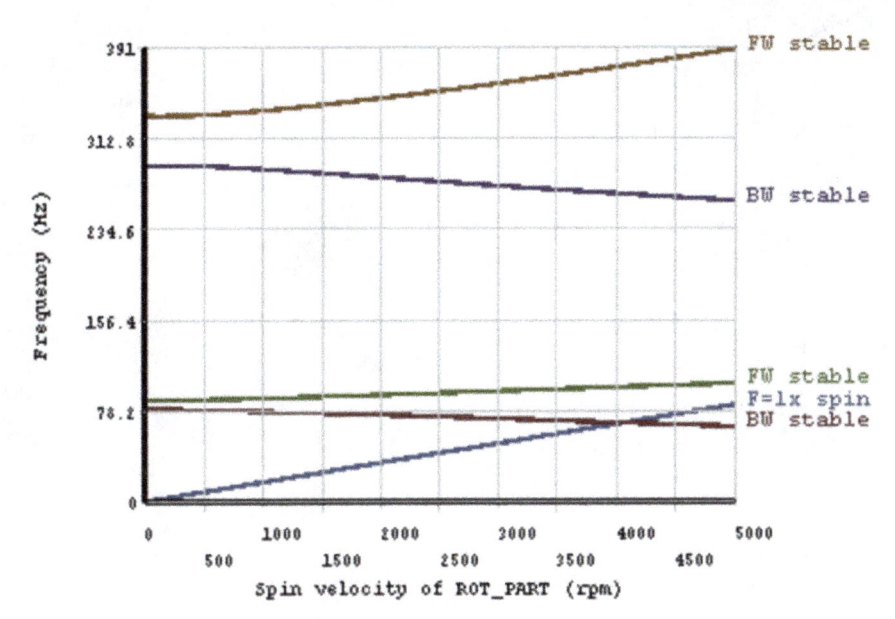

Fig. 16. Campbell Diagram for the propeller shaft

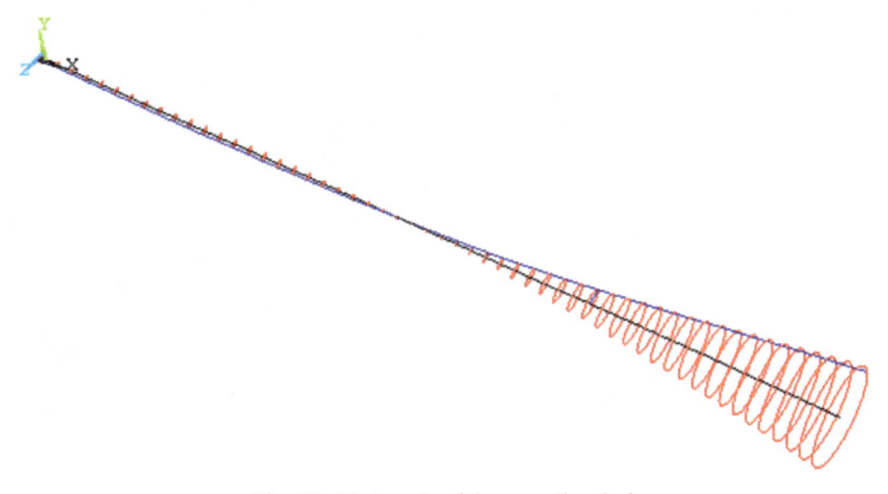

Fig. 17. First mode of the propeller shaft

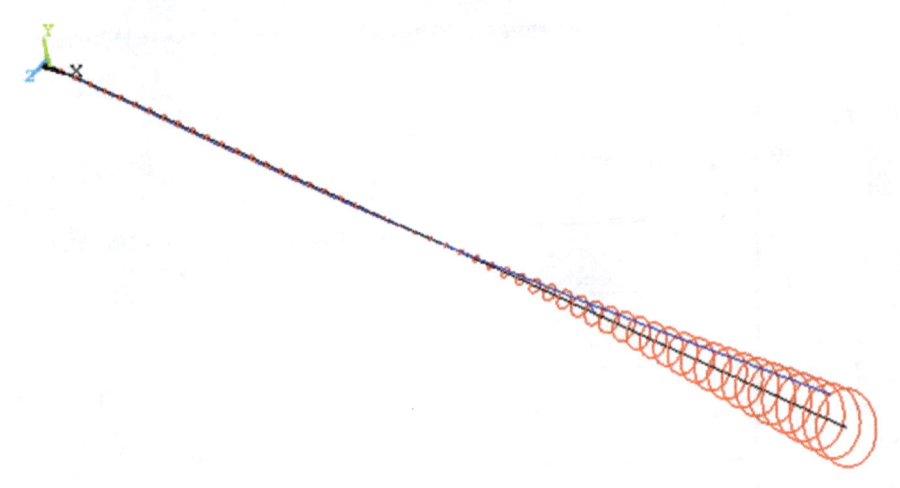

Fig. 18. Second mode shape of the propeller shaft

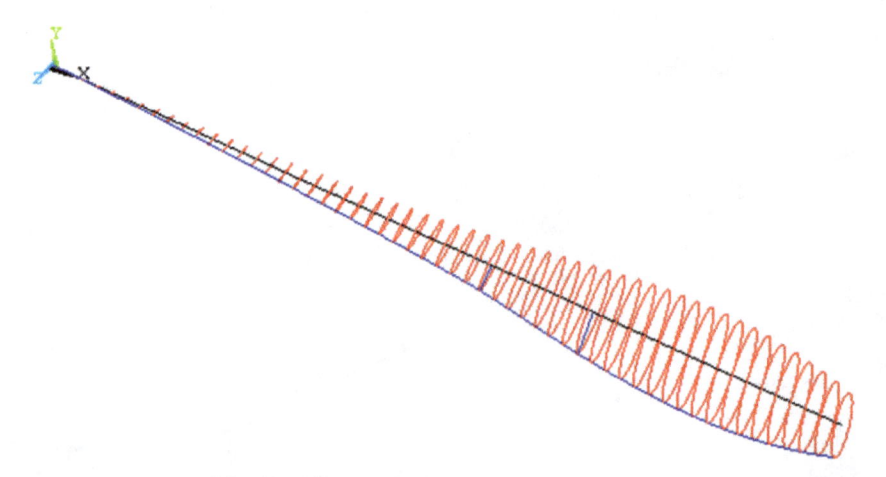

Fig. 19. Third mode shape of the propeller shaft

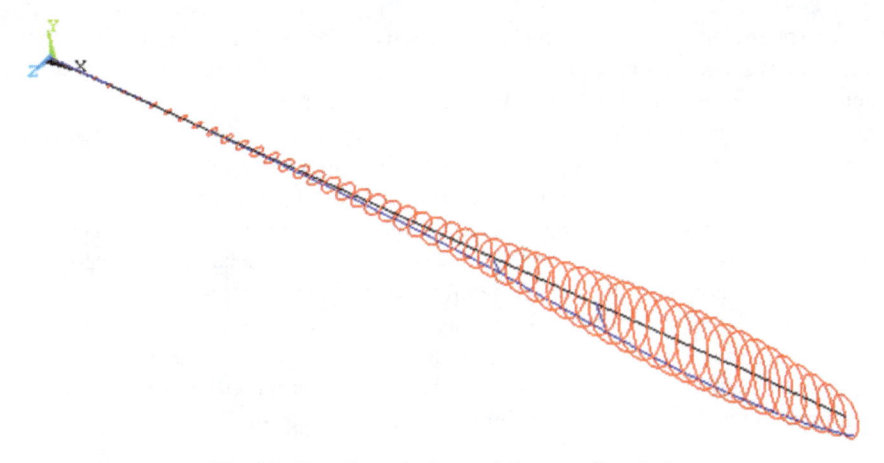

Fig. 20. Fourth mode shape of the propeller shaft

5 Conclusions and Future Work

The results of the simulations presented above, show an acceptable behavior of an AMB as a stern tube bearing. By selecting the appropriate values for the geometry parameters and relatively low current values, a sufficient load capacity could be succeeded, while a PID controller can stabilize the system and eliminate any undefined displacements. The linear force-current and force-displacement equation is validated for a range 0–0.6 A and for displacements smaller than 0.2 mm, which means that a feedback control system can be designed using the linearized model. The displacement and current stiffness factors are also calculated with FEM simulations. The controller's gains, calculated regarding the proposed stiffness and damping values from the literature, offer adequate control of the rotor's displacement for step disturbances up to 5 kN. This analysis also showed that critical speeds are far above the operational speed range. The shaft in the current analysis operates at 1940 rpm and the first critical speed is calculated at 4776 rpm.

However, the dynamics of the propeller shaft are a significant challenge for future work when it comes to an AMB. High-power density is the future development trend of these machines, which demands that the rotor is characterized as high-speed and slender, while operating above the critical speeds. Furthermore, it is a big challenge for a flexible rotor to pass the bending critical speeds and operate above those speeds steadily and reliably. The propeller shaft is an example of an overhung rotor with a heavy load (propeller) on its end. This leads to excessive shear stresses and gyroscopic loads because of the propeller's mass and inertia. While the present analysis focuses on the rigid rotor case, a flexible rotor case could be addressed with proper development of the feedback control system model. More specifically, the rigid rotor's block, used in this analysis, could be replaced with one that calculates rotor's displacements using either the Euler-Bernoulli, or the Timoshenko beam theory. This could be possible by combining a rotor dynamics FEM model with the MATLAB model of the feedback control. When it comes to a possible bending deformation of the rotor inside the AMB, a solution could be given by multiple short length AMBs placed in short distance to divide the load carried and control any angular displacements more effectively. As for the power losses, this analysis makes a first approach of the Ohmic losses on the coils, which can be significant when it comes to heavier loads and more powerful magnetic fields. At high rotational speeds and/or high magnetic flux density values though, magnetic losses are significant and need to be calculated and considered. When it comes to system control theory, the decoupled control case was presented, which is sufficient for a symmetric system. Once the gyroscopic phenomenon is considered though a coupled control system is needed, where the control of each direction takes the other directions behavior into account.

Each of the issues mentioned above create a challenge for the researchers of each field. Since efficiency of control system and electric machines is a hot research topic that has progressed significantly recently, it seems that the efficiency of the AMBs will increase furtherly in the near future and their application will expand, so that ship machinery could also benefit by this innovative type of bearing.

Acknowledgement. The present work was financially supported by the «Andreas Mentzelopoulos Foundation».

References

1. Sestan, D.S.: Experimental determination of stern tube journal bearing behaviour. Brodo Gradnja **61**(2), 130–141 (2010)
2. Maslen, E.H., Schweitzer, G.: Magentic Bearings: Theory, Design and Application to Rotating Machinery. Springer, Heidelberg (2009)
3. Chalam, R., Shelke, S.: Optimum energy loss in electromagnetic bearing. In: 3rd International Conference on Electronics Computer Technology (ICECT), Kanyakumari, India (2011)
4. Tomczuk, B., Wajnert, D., Zimon, J.: Field-circuit modeling of AMB system for various speeds of the rotor. J. Vibroengineering **14**(1), 165–170 (2012)
5. Braut, S., Bulić, N., Zigulic, R., Stimac, G.: Modeling and experimental verification of a flexible rotor/AMB system. COMPEL – Int. J. Comput. Math. Electr. Electron. Eng. **32**(4), 1244–1254 (2013)
6. Potawad, A., Pujari, M., Rane, R., Pewekar, M.M.: Analysis of Active Magnetic Bearings. Department of Mechanical Engineering, MCT's Rajiv Gandhi Institute of Technology, Mumbai (2018)
7. Nikolakopoulos, P., Papadopoulos, C., Farmakopoulos, M.G.: Design of an active hydromagnetic journal bearing. In: Proceedings of the 8th IFToMM International Conference on Rotordynamics, Seoul, Korea (2010)
8. Nikolakopoulos, P., Papadopoulos, C., Farmakopoulos, M.G.: Design of an active hydromagnetic journal bearing. J. Eng. Tribol. **227**(7), 673–694 (2012)
9. Bompos, D., Nikolakopoulos, P.G., Papadopoulos, C.A., Farmakopoulos, M.: The role of the AMB on the ship propulsion. In: 1st International MARINELIVE Conference on "All Electric Ship", Athens, Greece (2012)
10. Enemark, S., Santos, I.F.: Nonlinear dynamic behaviour of a rotor–foundation system coupled through passive magnetic bearings with magnetic anisotropy – theory and experiment. J. Sound Vibr. **363**, 407–427 (2016). https://doi.org/10.1016/j.jsv.2015.10.007
11. Lijesh, K.P., Muzakkir, S.M., Hirani, H.: Failure mode and effect analysis of passive magnetic bearing. J. Eng. Fail. Anal. **62**, 1–20 (2016)

12. Sun, J.J., Wang, C., Le, Y.: Research on a novel high stiffness axial passive magnetic bearing for DGMSCMG. J. Magn. Magn. Mater. **412**, 147–155 (2016)
13. Fang, J.C., Zheng, S.Q., Han, B.C.: Attitide sesnsing and dybamic decoupling based on active magnetic bearing of MSDGCMG. J. IEEE Trans. Instrum. Measur. **61**, 338–348 (2012)
14. Cui, P., He, J., Fang, J.C.: Static mass imbalance identification and vibration control for rotor of magentically suspended control moment gyro with active-passive magnetic bearings. J. Vib. Control **22**, 2313–2324 (2016)
15. Smirnov, A., Uzhegov, N., Sillanpää, T., Pyrhonen, J., Pyrhonen, O.: High-speed electrical machine with active magentic bearing system optimization. IEEE Trans. Ind. Electron. **64**, 9876–9885 (2017)
16. Schuck, M., Steinert, D., Kolar, J.W.: Active radial magnetic bearing for an ultra-high speed motor. In: 18th European Conference on Power Electronics and Applications (EPE 2016 ECCE Europe) (2016)
17. Sun, J.J., Zhou, H., Ma, X., Ju, Z.Y.: Study on PID tuning strategy based on dynamic stiffness for radial active magnetic bearing. J. ISA Trans. **80**, 458–474 (2018)
18. Sun, J.J., Ren, Y., Fang, J.C.: Passive axial magnetic bearing with Halbach magnetized array in magnetically suspended control moment gyro application. J. Magn. Magn. Mater. **323**, 2103–2107 (2011)
19. Yang, J.Y., Zhang, W.Y., Ge, Y.J.: A precise measurement of windings axes and stiffness coefficient for power magnetic bearing. In: International Conference on Electrical and Control Engineering (ICECE) (2010)
20. Cavalca, K.L., da Silva Tuckmantel, F.W.: Vibration signatures of a roter-coupling-bearing system under angular misalignment. Mech. Mach. Theory **133**, 559–583 (2019)
21. Castro, H., Cavalva, K.L., Tuckmantel, F.: Investigation on vibration response for misaligned rotor-bearing-flexible disc coupling system-theory and experiment. J. Vib. Acoust. **142** (2020)
22. Tiwari, R., Kumar, P.: Finite element modelling, analysis and identification using novel trial misalignment approach in an unbalanced and misaligned flexible rotor system levitated by active magnetic bearings. Mech. Syst. Sig. Process **151** (2021)
23. Dimitri, A.S., Maslen, E.H., El-Shafei, A., Shaltout, M.L.: Control of a smart electromechanical acuator journal integrated bearing to a common equilibrium position: a simulation study. Mech. Syst. Signal Process. **154** (2021)
24. Metals Handbook, 8th edn, vol. 1. American Society for Metals (ASM) (1961)
25. Antila, M.: Electromechanical Properties of Radial Active Magnetic Bearings. Finnish Academy of Technology, Helsinki (1998)
26. ANSYS Help Documentation. https://ansyshelp.ansys.com. Accessed Feb 2022
27. Secara, E., Mihalcica, M., Dumitru, N.: Study of rotor-bearing systems using Campbell diagram. In: Proceedings of the 1st International Conference on Manufacturing Engineering, Quality and Production Systems, vol. II, Brasov, Romania (2009)
28. Someya, T. (ed.): Journal-Bearing Data-Book. Springer, Heidelberg (1989)

On the Stability Margins of Parametrically Excited Rotating Shafts on Gas Foil Bearings: Linear and Nonlinear Approach

Emmanouil Dimou[1], Fadi Dohnal[2], and Athanasios Chasalevris[1(✉)]

[1] National Technical University of Athens, Athens 15780, Hellas
chasalevris@mail.ntua.gr
[2] Vorarlberg University of Applied Sciences, 6850 Dornbirn, Austria

Abstract. Parametric excitation is applied in a realistic multi-segmented rotating shaft of a turbopump mounted on two active gas foil bearings. The active configuration of the two gas foil bearings is defined by a periodic external load of specific amplitude and frequency which alternates the top foil configuration and in this manner the gas film impedance forces experience periodic variation.

The analytical model of the rotor is obtained using a reduced finite element model, and the Reynolds equation for the compressible flow of the gas is solved applying a reduced finite difference scheme.

Fully balanced rotors are investigated on their potential to shift the threshold speed of instability defined by the rotating speed at which the first bifurcation of limit cycles occurs. The limit cycles are evaluated through pseudo arc length continuation with an embedded collocation method. The respective amplitude and frequency of applied external excitation (parametric excitation) is investigated in order to define those characteristics (amplitude and frequency) which render parametric antiresonance in the rotating system. Two approaches are included. At the first, the parametric excitation is implemented in the system, through periodically varying stiffness and damping coefficients of the gas foil bearings, which are evaluated solving the perturbed Reynolds equation; this is the linear version of the dynamic system. At the second, direct implementation of gas bearing impedance forces is considered; this is the nonlinear version of the dynamic system. Comparing the two approaches, it is found that antiresonance occurs in specific excitation frequencies, and the rotating system operates without bifurcation at speeds two times higher than the respective speed of the reference system without excitation. Several design scenarios for rotor slenderness and bearing configuration are included in the results.

Keywords: Parametric Excitation · Rotating Shafts · Gas Foil Bearings · Antiresonance

1 Introduction

Parametrically excited systems (systems with periodically changing physical properties) with multiple degrees of freedom have been extensively studied in both mathematical and engineering applications [1–3]. A well-chosen parametric excitation couples certain modes and allows a more efficient usage of their existing damping [4, 5]. Parameter regions where trivial solutions of unstable systems retain their stability under the effect of parametric excitation are called parametric antiresonances. In current work, parametric excitation has been investigated on its potential to introduce parametric antiresonances and therefore extent the stability margins of a turbopump rotor mounted on gas foil bearings.

One of the first attempts to implement parametric excitation in structural systems has been done in [6], where a uniform cantilever beam under the effect of a periodic axial load was investigated. Two realistic rotor models were investigated in [7] in order to prove that an initially unstable equilibrium position can be stabilized by introducing a parametric excitation. Another realistic rotor system with adjustable oil film journal bearings was investigated in [8], where the main goal was the enhancement of the stability margins of a turbine rotor modeled with FEM. The works hereby referred do not consider complex rotor models coupled to gas foil bearing models; active gas foil bearings as a mean to introduce parametric excitation in a rotor system are investigated in this paper for first time. Additionally, numerical continuation has been lately applied mostly in simplistic nonlinear rotor-bearing systems. Among various contributions, in [9–11] simplified models of high speed rotors on floating ring bearings are investigated, while in [12–15] the bifurcation set of Jeffcott rotor model coupled to simple oil film bearing model was studied; no parametric excitation was considered in these studies. Recently, the bifurcation sets of simple rotor models on adjustable oil bearings [16] and on gas-foil bearings [17] where studied. The works hereby referred do not consider complex rotor models mounted on parametrically excited gas foil bearings as most of the effort was on the respective bearing models.

In this work, parametric excitation is introduced by a predefined sinusoidal displacement (of specific amplitude and frequency) of the outer deformable ring of the gas bearing, which can be achieved using piezo-actuators [18]. In order to approximately determine the excitation frequencies of great interest (the so-called fundamental parametric antiresonance frequencies) the simplistic approach of the linearized stiffness and damping coefficients [19, 20] is adopted. The aforementioned approach is accurate when the journal's orbit is sufficiently close to an equilibrium position (fixed point). Therefore, after determining the excitation frequencies of interest, a nonlinear approach of the elastoaerodynamic lubrication problem is adopted too. The fluid- structural model of the parametrically excited gas foil bearing is coupled to the rotor's dynamic MDoF model. A model order reduction for the fluid equation flow (Reynolds equation) is implemented in order to obtain an overall bearing model suitable for fully coupled nonlinear rotor-dynamic investigations [21].

The solution branches of the parametrically excited and perfectly balanced rotor – bearing system as one operating parameter (the rotating speed) changes, are evaluated via the most reputable continuation method [22–25], the pseudo arc length continuation method. This method has the primary advance to study MDoF systems, where the nonlinear equations of motion can be many [26]. It incorporates a collocation method for finding periodic solutions of both linear and nonlinear first order ODEs, and in this paper it is programmed by the authors, directly from the notes [22, 25 and 27]. Only perfectly

190 E. Dimou et al.

balanced rotors are considered in this study in order to avoid quasi-periodic motions of the rotor, resulted by the simultaneous parametric and synchronous (unbalance) excitation.

2 Model of a Rotor on Active Gas Bearings with Parametric Excitation

2.1 Elastoaerodynamic Lubrication and Resulting Gas Forces

The assumptions introduced in the gas-lubrication problem are: a) isothermal gas film, b) laminar flow, c) no-slip boundary conditions, d) continuum flow, e) negligible fluid inertia, f) ideal isothermal gas law $(p/\rho = ct.)$, , g) negligible entrance and exit effects, and negligible curvature of the gas film as a result of small bearing clearance $(R \approx R + c_r)$. The Reynolds equation for compressible gas flow is given in Eq. (1) in dimensionless form, and it is an implicit function of time and of journal and foil kinematics.

$$\frac{\partial}{\partial \bar{x}}\left(\bar{p}\bar{h}^3 \frac{\partial \bar{p}}{\partial \bar{x}}\right) + \kappa^2 \frac{\partial}{\partial \bar{z}}\left(\bar{p}\bar{h}^3 \frac{\partial \bar{p}}{\partial \bar{z}}\right) = \bar{\Omega}\frac{\partial}{\partial \bar{x}}\left(\bar{p}\bar{h}\right) + 2\frac{\partial}{\partial \tau}\left(\bar{p}\bar{h}\right) \tag{1}$$

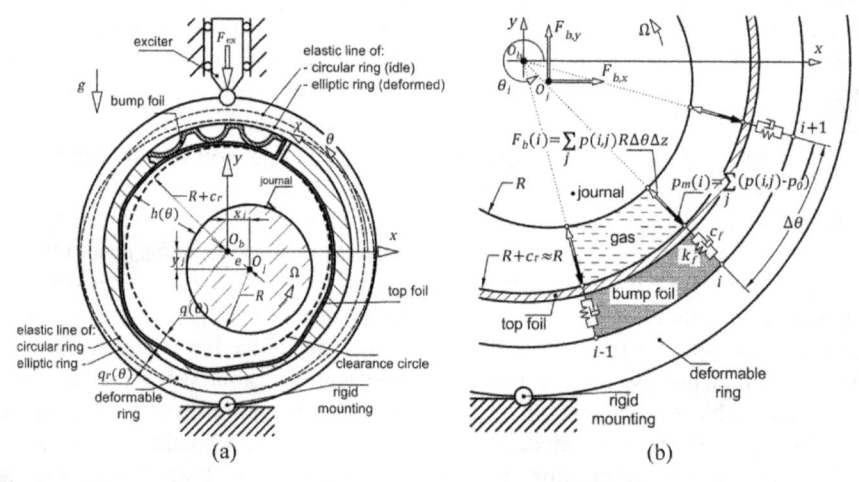

Fig. 1. (a) Representation and key design properties of the gas foil bearing under the effect of parametric excitation force acting at the outer ring, (b) modeling of the bump foil and the respective forces acting on the components of the gas foil bearing.

Analytical solution for Eq. (1) cannot be defined, and an indicated approach to evaluate the pressure distribution is the Finite Difference Method - FDM. The pressure domain is converted into a grid of $i = 1, \ldots, N_{\bar{x}} + 1$ and $j = 1, \ldots, N_z + 1$ mesh points (i, j are the indexes in the circumferential and axial direction, see Fig. 1).

The Reynolds equation is first rewritten defining the first-time derivative of the pressure in Eq. (2). Then the discrete Reynolds equation is defined in the grid points expressing the partial derivatives with finite differences.

$$\dot{\bar{p}} = \frac{1}{2\bar{h}}\frac{\partial}{\partial \bar{x}}\left(\bar{p}\bar{h}^3\frac{\partial \bar{p}}{\partial \bar{x}}\right) + \frac{\kappa}{2\bar{h}}\frac{\partial}{\partial \bar{z}}\left(\bar{p}\bar{h}^3\frac{\partial \bar{p}}{\partial \bar{z}}\right) - \frac{\bar{\Omega}}{2\bar{h}}\frac{\partial}{\partial \bar{x}}\left(\bar{p}\bar{h}\right) - \frac{\bar{p}\dot{\bar{h}}}{\bar{h}} \tag{2}$$

The dimensionless parameters of gas pressure $\bar{p}$, gas film thickness $\bar{h}$, , spatial coordinates (circumferential and axial respectively) $\bar{x} = \theta$ and $\bar{z}$, dimensionless time τ, dimensionless rotating speed $\bar{\Omega}$, and ratio $\kappa = R/L_b$ are included in the elastoaerodynamic lubrication problem of Eq. (2). The gas film thickness is defined in Eq. (3) for the continuous and the discrete pressure domain (finite difference grid) where $\bar{q} = \bar{q}(\theta)$ (or $\bar{q}_i = \bar{q}(\theta_i)$ in the discrete pressure domain) is the foil deformation in radial direction, see Fig. 1.

$$\bar{h} = 1 - \bar{x}_j \cos\theta - \bar{y}_j \sin\theta + \bar{q}, \quad \bar{h}_i = 1 - \bar{x}_j \cos\theta_i - \bar{y}_j \sin\theta_i + \bar{q}_i \qquad (3)$$

Boundary conditions of the elastoaerodynamic lubrication problem are defined both in continuous and discrete domain below (the symmetry of the problem in the axial direction has been taken into account).

$$\bar{p}(\tau, \theta_0, \bar{z}) = \bar{p}(\tau, \theta_0 + 2\pi, \bar{z}) = 1, \quad \bar{p}_{1,j} = \bar{p}_{Nx+1,j} = 1$$
$$\frac{\partial \bar{p}}{\partial \bar{z}}|_{\bar{z}=1/2} = 0, \quad \frac{\bar{p}_{i,Nz/2} - \bar{p}_{i,Nz/2-1}}{\Delta \bar{z}} = 0 \qquad (4)$$

It is important to mention that a common assumption in GFBs is sub-ambient pressures to arise. The sub-ambient pressure can cause the top foil to separate from the bump foil into a position in which the pressure on both sides of the pad is equalized. Heshmat [8] first introduced a set of boundary conditions accounting for this separation effect. More specifically, a Gümbel boundary condition is imposed, meaning that sub-ambient pressure is discarded when integrating the pressure distribution over the bearing's length, in order to obtain the bearing force components. In terms of numerical calculations, the assumption made by Heshmat can be simply explained. In case fluid pressure p is lower than the ambient p_0, then the former should be considered equal to p_0 and the foil deformation at these points will be zero ($\bar{q}_i = 0$ for $\bar{p}_i < 1$).

The simplified model for the bump foil structure is depicted at Fig. 1. The structure consists of $N_{\bar{x}} - 2$ linear massless elements of stiffness $\bar{k}_f$ (compliance $\bar{a}_f = 1/\bar{k}_f$) and damping $\bar{c}_f$. The springs and dampers mount the corresponding $N_{\bar{x}} - 1$ top foil stripes of area $\Delta \bar{x} \cdot L_b$ (or dimensionless area $\Delta \bar{x} \cdot 1$). The top foil of the bearing does not cover a complete cylinder; a single gap can be found at $\theta = \theta_0$, where foil is clamped to the bearing housing. Therefore, the moving top foil stripes are $N_{\bar{x}} - 1$. The top foil stripes are assumed to remain parallel to the bearing surface during their motion, therefore, no axial coordinate is required for the top foil motion. The motion of each of the top foil stripe is excited by the mean gas pressure $\bar{p}_{m,i}$ acting on the top of it, creating the gas force $\bar{F}_B(i)$. The mean gas pressure $\bar{p}_{m,i}$ is defined in Eq. (5) (in the continuous and discrete pressure domain), in dimensional and dimensionless form.

$$p_m(\theta) = \frac{1}{L_b} \int_0^{L_b} p(\theta) dz, \quad p_{m,i} = \frac{1}{L_b} \sum_{j=2}^{Nz} (p_{i,j} \Delta z)$$
$$\bar{p}_{m,i} = \sum_{j=2}^{Nz} (\bar{p}_{i,j} \Delta \bar{z}) = \frac{1}{Nz} \sum_{j=2}^{Nz} (\bar{p}_{i,j}) \qquad (5)$$

The foil stiffness and damping coefficient are defined as $\bar{k}_f = 1/\bar{a}_f$ and $\bar{c}_f = \eta \cdot \bar{k}_f$ for foil motion synchronous to the pressure excitation (η denotes the loss factor); the simplistic assumption follows the work [28]. The $N_{\bar{x}} - 1$ ODEs describing the top foil radial displacement $\bar{q}_i$ of the stripe i under the effect of a periodically changing outer ring radial deformation $\bar{\mathbf{q}}_r = \{\bar{q}_{r,i}\}$ and the pressure excitation are defined in Eq. (6).

$$\dot{\bar{q}}_i = \dot{\bar{q}}_{r,i} + \frac{[\bar{p}_{m,i} - \bar{k}_f(\bar{q}_i - \bar{q}_{r,i})]}{\bar{c}_f}, \quad i = 2, 3, \ldots, N_x \tag{6}$$

Finally, the nonlinear gas forces, after evaluating pressure distribution $\bar{p}$ (as $\bar{p}_{i,j}$), can be evaluated according to Eq. (7) where $\Delta \bar{x} = 2\pi/N_{\bar{x}}$, $\Delta \bar{z} = 1/N_z$.

$$\begin{aligned}
\bar{F}_x^B &= -\int_0^{2\pi}\int_0^1 (\bar{p} - 1)\cos\theta \, d\theta \, d\bar{z} = -\sum_{i=2}^{N_x}\sum_{j=2}^{N_z}(\bar{p}_{i,j} - 1)\cos\theta_i \, \Delta\bar{x}\,\Delta\bar{z} \\
\bar{F}_y^B &= -\int_0^{2\pi}\int_0^1 (\bar{p} - 1)\sin\theta \, d\theta \, d\bar{z} = -\sum_{i=2}^{N_x}\sum_{j=2}^{N_z}(\bar{p}_{i,j} - 1)\sin\theta_i \, \Delta\bar{x}\,\Delta\bar{z}
\end{aligned} \tag{7}$$

The linearized gas forces (useful only for a rough approximation of the fundamental excitation frequency) can be calculated after the evaluation of the linearized stiffness and damping coefficients. The perturbation method following is explained. Supposing that the perturbed motion of the journal about an equilibrium position (subscript '0' indicates a quantity in this equilibrium position) is a circular orbit, the normalized perturbations (small displacements and small velocities) are given by Eq. (8).

$$\Delta \bar{X}(\tau) = \frac{\Delta X(\tau)}{c_r} = |\Delta \bar{X}|e^{i\tau}, \quad \Delta \bar{Y}(\tau) = \frac{\Delta Y(\tau)}{c_r} = |\Delta \bar{Y}|e^{i\tau} \tag{8}$$

Then, the pressure distribution, the foil deformation and the film thickness can be expressed in Taylor series in terms of the normalized perturbations around an equilibrium position of the journal in Eq. (9).

$$\begin{aligned}
\bar{p}_{i,j} &= \bar{p}_{0,i,j} + \bar{p}_{\bar{X},i,j}\Delta\bar{X} + \bar{p}_{\bar{Y},i,j}\Delta\bar{Y} + \bar{p}_{\dot{\bar{X}},i,j}\Delta\dot{\bar{X}} + \bar{p}_{\dot{\bar{Y}},i,j}\Delta\dot{\bar{Y}} \\
\bar{h}_i &= \bar{h}_{0,i} + \bar{h}_{\bar{X},i}\Delta\bar{X} + \bar{h}_{\bar{Y},i}\Delta\bar{Y} + \bar{h}_{\dot{\bar{X}},i}\Delta\dot{\bar{X}} + \bar{h}_{\dot{\bar{Y}},i}\Delta\dot{\bar{Y}} \\
\bar{q}_i &= \bar{q}_{0,i} + \bar{q}_{\bar{X},i}\Delta\bar{X} + \bar{q}_{\bar{Y},i}\Delta\bar{Y} + \bar{q}_{\dot{\bar{X}},i}\Delta\dot{\bar{X}} + \bar{q}_{\dot{\bar{Y}},i}\Delta\dot{\bar{Y}}
\end{aligned} \tag{9}$$

The relationships between the partial derivatives of the film thickness and the foil deformations are given by:

$$\begin{aligned}
\bar{h}_{\bar{X},i} &= \bar{q}_{\bar{X},i} + \sin\theta_i, \quad \bar{h}_{\dot{\bar{X}},i} = \bar{q}_{\dot{\bar{X}},i} \\
\bar{h}_{\bar{Y},i} &= \bar{q}_{\bar{Y},i} - \cos\theta_i, \quad \bar{h}_{\dot{\bar{Y}},i} = \bar{q}_{\dot{\bar{Y}},i}
\end{aligned} \tag{10}$$

Equations (8)–(10) are substituted in Eq. (1). Terms with zero subscript are collected yielding to the equations about the journal's equilibrium position. After the evaluation of the journal's equilibrium position, the perturbed pressure distribution can also be evaluated by collecting all the other terms with the same coefficients ($\Delta\bar{X}$, $\Delta\bar{Y}$, $\Delta\dot{\bar{X}}$, $\Delta\dot{\bar{Y}}$). Finally, the linearized stiffness and damping coefficients

$\overline{K}_{xx}, \overline{K}_{xy}, \overline{K}_{yx}, \overline{K}_{yy}, \overline{C}_{xx}, \overline{C}_{xy}, \overline{C}_{yx}, \overline{C}_{yy}$ are evaluated integrating the perturbed pressure distributions over the bearing's surface and the linearized bearing forces are given in Eq. (11). For more detailed information, the reader may refer to [18].

$$\left\{\begin{array}{c}F_x^B \\ F_y^B\end{array}\right\} = \left\{\begin{array}{c}F_x^B \\ F_y^B\end{array}\right\}_0 + \left\{\begin{array}{c}\overline{K}_{xx} \\ \overline{K}_{yx}\end{array}\right\} \cdot \Delta \overline{X} + \left\{\begin{array}{c}\overline{K}_{xy} \\ \overline{K}_{yy}\end{array}\right\} \cdot \Delta \overline{Y} + \left\{\begin{array}{c}\overline{C}_{xx} \\ \overline{C}_{yx}\end{array}\right\} \cdot \Delta \dot{\overline{X}} + \left\{\begin{array}{c}\overline{C}_{xy} \\ \overline{C}_{yy}\end{array}\right\} \cdot \Delta \dot{\overline{Y}}$$
(11)

2.2 Condensed Rotor Model

The rotor system studied in current work is a representative turbopump rotor operating in speeds higher than 20 kRPM, mounted on two identical active gas foil bearings (AGFBs), see Fig. 2.

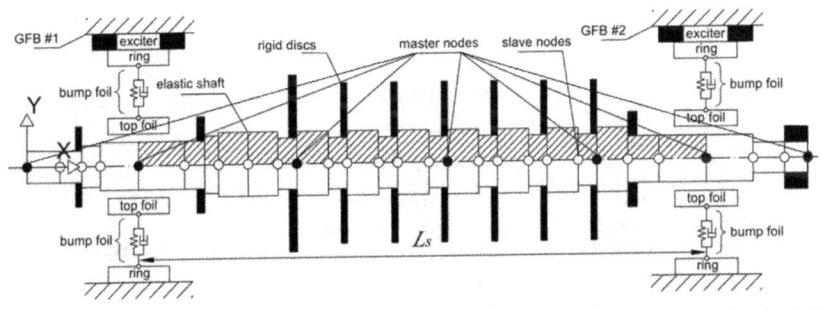

Fig. 2. Schematic representation of a slender high-speed rotor supported on two identical GFBs. Finite element discretization, bearing span L_s, and master and slave nodes are also depicted.

The rotor has complex geometry with different material properties and additional masses in various locations. Temperature distribution is additionally taken into account. The rotor is discretized with finite (beam) elements, with each element having two nodes and a total of eight degrees of freedom x_i (two transverse displacements and two tilting angles per node). The global inertia, gyroscopic and stiffness matrices are assembled by the proper summation of the individual finite beam element matrices. The global damping matrix is given by the classical Rayleigh formula. Thus, the equations of motion for the whole rotor system are derived in Eq. (12).

$$[\mathbf{M}]\{\ddot{\overline{x}}_i\} + ([\mathbf{C}] + [\mathbf{G}])\{\dot{\overline{x}}_i\} + [\mathbf{K}]\{\overline{x}_i\} = \left\{\overline{F}_i^B\right\} + \left\{\overline{F}_i^G\right\}$$
(12)

The rotor is considered perfectly balanced and unbalance forces are not included in the right-hand side of the equations of motion. If the bearing forces $\left\{\overline{F}_i^B\right\}$ are evaluated according to the nonlinear elastoaerodynamic approach, then they are the single source of nonlinearity in the system (as all of them are implicit functions of the journals' kinematics), and the equations of motion are nonlinear ODEs. The bearing forces can alternatively be evaluated using the periodically changing linearized stiffness and damping coefficients; then, equations of motion are linear ODEs with time periodic coefficients. In order include the gravity forces $\{\overline{F}_i^G\}$, it is supposed that the mass of each element is equally divided to the two nodes of the element.

The rotor model is reduced using the static (Guyan) reduction method. Only the transverse displacements of each master node are retained. The total number of master nodes is 7 thus 14 master DoFs are included in the vector $\{\overline{x}_{m,i}\}$. The selection of master nodes has been performed in order to match the dynamic response of the reduced system, see Eq. (13) to this of the full system (see Eq. (12)) in terms of unbalance response, and modal properties. The aforementioned performance is validated among others, in Sect. 3.

$$[\mathbf{M}_r]\{\ddot{\overline{x}}_{m,i}\} + ([\overline{\mathbf{C}}_r] + [\mathbf{G}_r])\{\dot{\overline{x}}_{m,i}\} + [\mathbf{K}_r]\{\overline{x}_{m,i}\} = \{\overline{F}_{r,i}^B\} + \{\overline{F}_{r,i}^G\} \quad (13)$$

The reduced rotor model equations of motion are then converted to a first order ODE set applying the transformation of Eq. (14).

$$\overline{\mathbf{y}}_m = \{\overline{y}_{m,i}\} = \begin{cases} \{\dot{\overline{x}}_{m,i}\} \\ \{\overline{x}_{m,i}\} \end{cases} \quad (14)$$

Therefore, the reduced rotor model equations of motion can be rewritten in Eq. (15) where $\{\overline{F}_{r,i}\} = \{\overline{F}_{r,i}^B\} + \{\overline{F}_{r,i}^G\}$.

$$\{\dot{\overline{y}}_{m,i}\}_{28\times 1} = \begin{bmatrix} \mathbf{0}_{14\times 14} & \mathbf{I}_{14\times 14} \\ (-\mathbf{M}_r^{-1}\mathbf{K}_r)_{14\times 14} & -\mathbf{M}_r^{-1}(\mathbf{G}_r + \overline{\mathbf{C}}_r)_{14\times 14} \end{bmatrix} \{\overline{y}_{m,i}\}_{28\times 1} + \begin{cases} \mathbf{0}_{14\times 1} \\ \mathbf{M}_r^{-1}\{\overline{F}_{r,i}\}_{14\times 1} \end{cases} \quad (15)$$

Due to the periodic deformation of the outer ring (or the periodically changing stiffness and damping coefficients), dimensionless time still appears explicitly in the equations of motion in Eq. (15) and the first order ODE system is non-autonomous, see Appendix.

2.3 Composition and Solution of the Dynamic System

At the case that the dynamic system is considered nonlinear, the gas bearing impedance forces should be evaluated at every discrete time moment, and the Reynolds equation should be reduced in size applying an order reduction method [9], improving the computational costs. This method can be categorized within the theory of weighted residuals and assumes that the pressure distribution can be approximated in Eq. (16) [9].

$$\overline{p}(\tau, \overline{x}, \overline{z}) \approx \overline{p}_a(\tau, \overline{x}, \overline{z}) + 1 = \hat{p}_a(\tau, \overline{x}) \cdot \tilde{p}_a(\overline{z}) + 1 \quad (16)$$

Obviously $\tilde{p}_a(\overline{z})$ should be treated as base function. Authors in [9] were motivated by the short bearing theory for journal bearings in order to choose the proper base function. In the current work, the same generalized shape function has been introduced, as in Eq. (17).

$$\tilde{p}_a(\overline{z}) = (1 - w_a) \cdot \left(1 - 2\overline{z}^{2i_a}\right) + w_a \cdot \left(1 - 2\overline{z}^{2i_a+2}\right)$$
$$i_a = floor(m_a), \ w_a = m_a - i_a, \ m_a \geq 1, m_a \in \mathbb{R} \quad (17)$$

The optimal parameter m_a can be determined by introducing an error function which includes the difference between the pressure field of the full domain model and the pressure field of the reduced one. The optimal parameter minimizes the aforementioned error function. For more detailed information, the reader may refer to [9]. In this way the aerodynamic problem renders $N_x - 1$ 1st order ODEs with respect to the time derivative of the nodal pressure, see Eq. (18).

$$\dot{\bar{\mathbf{p}}} = \{\dot{\bar{p}}_{a,i}\} = \mathbf{f}_B(\bar{\mathbf{p}}, \bar{\mathbf{q}}, \bar{\mathbf{q}}_r, \bar{\mathbf{y}}_m) \tag{18}$$

The $N_x - 1$ 1st order ODEs which describe the top foil's radial displacement, see Eq. (6), are written in Eq. (19).

$$\dot{\bar{\mathbf{q}}} = \{\dot{\bar{q}}_i\} = \mathbf{f}_F(\bar{\mathbf{p}}, \bar{\mathbf{q}}, \bar{\mathbf{q}}_r) \tag{19}$$

The reduced rotor model equations of motion, see Eq. (15), are written in Eq. (20).

$$\dot{\bar{\mathbf{y}}}_m = \mathbf{f}_R(\bar{\mathbf{p}}, \bar{\mathbf{y}}_m) \tag{20}$$

At the case that the dynamic system is considered linear, then the bearing forces $\{\bar{F}^B_{r,i}\}$ should be evaluated using the parametrically excited linearized stiffness and damping coefficients in Eq. (21), where $\bar{\Omega}_{ex}$ denotes the dimensionless parametric excitation frequency and δ denotes the dimensionless amplitude.

$$\begin{bmatrix} \bar{K}_{xx} & \bar{K}_{xy} \\ \bar{K}_{yx} & \bar{K}_{yy} \end{bmatrix}^{p.e} = \begin{bmatrix} \bar{K}_{xx} & \bar{K}_{xy} \\ \bar{K}_{yx} & \bar{K}_{yy} \end{bmatrix} (1 + \delta \sin(\bar{\Omega}_{ex}\tau))$$
$$\begin{bmatrix} \bar{C}_{xx} & \bar{C}_{xy} \\ \bar{C}_{yx} & \bar{C}_{yy} \end{bmatrix}^{p.e.} = \begin{bmatrix} \bar{C}_{xx} & \bar{C}_{xy} \\ \bar{C}_{yx} & \bar{C}_{yy} \end{bmatrix} (1 + \delta \sin(\bar{\Omega}_{ex}\tau)) \tag{21}$$

Note that the stiffness and damping properties of both bearings vary at the same frequency and phase. Equation (21) provides the ability to determine excitation frequency and amplitude which leads to parametric anti-resonances. These desirable parameters are then attempted to be achieved by the time dependent deformation of the outer ring and by using the nonlinear model, which is considered to be the realistic case.

At both versions of the dynamic system (linear or nonlinear), the rotor-gas foil bearing model is described by a non-autonomous 1st order ODE system in Eq. (22).

$$\dot{\bar{\mathbf{s}}} = \mathbf{f}(\bar{\mathbf{s}}, \bar{\Omega}, \bar{\Omega}_{ex}, \tau) \tag{22}$$

The aforementioned system (of dimension N) has to be converted to autonomous in order limit cycle solutions to be evaluated by collocation method [10–13]. This can be achieved by coupling Eq. (22) to a two DoF oscillator, see Eq. (23), whose unique solution is a harmonic function of frequency $\bar{\Omega}_{ex}$, as $\bar{s}_{N+1} = \cos(\bar{\Omega}_{ex}\tau)$ and $\bar{s}_{N+2} = \sin(\bar{\Omega}_{ex}\tau)$.

$$\dot{\bar{s}}_{N+1} = f_{N+1} = \bar{s}_{N+1} + \bar{\Omega}_{ex}\bar{s}_{N+2} - \bar{s}_{N+1}\left(\bar{s}^2_{N+1} + \bar{s}^2_{N+2}\right)$$
$$\dot{\bar{s}}_{N+2} = f_{N+2} = -\bar{\Omega}_{ex}\bar{s}_{N+1} + \bar{s}_{N+2} - \bar{s}_{N+2}\left(\bar{s}^2_{N+1} + \bar{s}^2_{N+2}\right) \tag{23}$$

The final (autonomous) 1st order ODE system (of dimension $N+2$) is then defined in Eq. (24), where $\dot{\tilde{\mathbf{s}}} = \{\bar{\mathbf{s}}^T \; s_{N+1} \; s_{N+2}\}^T$ and $\tilde{\mathbf{f}} = \{\mathbf{f}^T \; f_{N+1} \; f_{N+2}\}^T$.

$$\dot{\tilde{\mathbf{s}}} = \tilde{\mathbf{f}}(\tilde{\mathbf{s}}, \bar{\Omega}) \qquad (24)$$

It should be noted that at the case of the nonlinear version, the two DoF oscillators are included in the definition of the periodic deformation of the outer ring, see Eq. (30). At the case of the linear version, the two DoF oscillators are included in the parametrically excited stiffness and damping coefficients, see Eq. (21). An orthogonal collocation method is then applied for the computation of periodic limit cycles produced by the 1st order ODE system in Eq. (24), at a constant $\bar{\Omega}_{ex}$. Numerical continuation of periodic limit cycles is finally applied for the evaluation of periodic solution branches. For more detailed information, the reader may refer to [10, 13]. Limit cycles are not always periodic though. At these cases, where quasi periodic limit cycles appear (or chaotic motions are produced), the current solution scheme is not adequate, and solution branches of quasi periodic or chaotic motions are not included in the results.

3 Results and Discussion

At first, the validity of the method for the calculation of the stiffness and damping coefficients for a GFB is proved. The bearing properties as well as the number of mesh points in both the circumferential and axial direction are presented in Table 1. It should be noted that the aforementioned discretization was selected according to a sensitivity analysis based on the integration of the pressure distribution over the bearing surface.

Table 1. Gas foil bearing's physical properties and number of mesh points

Parameters	Values
Ambient pressure, $p_0[Pa]$	10^5
Gas dynamic viscosity, $\mu[mPa \cdot s]$	0.018
Radius of the bearing, $R[mm]$	15
Length of the bearing, $L_b[mm]$	30
Clearance, $c_r[\mu m]$	31
Starting/ending angle, $\chi[rad]$	$\pi/2$
Mesh intervals in circ. Direction, N_x	43
Mesh intervals in axial Direction, N_z	23

In Fig. 3, the predictions of the direct and cross coupling stiffness and damping coefficient are depicted for three different cases of dimensionless compliance $\bar{a}_f$ and for three different values of the bearing number $\Lambda = 6\mu R^2 \Omega/(p_0 c_r^2)$. It should be noted that according to J. P. Peng and M. Carpino [18], zero compliance of the bump foil is not identical to the rigid foil, as in the former case the top foil is not losing contact to the foundation when sub-ambient pressure occurs. It can be observed that the predictions show very good agreement if the foil is not assumed rigid. Actually, in case of zero compliance (or according to J. P. Peng and M. Carpino in case of rigid foil) slight discrepancies are observed and a possible explanation can be found in the numerous differences between the current method and that proposed by them.

In general, the resulting stiffness coefficient of the GFB is decreased as the bump foil compliance is increased, due to the increased deflection of the top foil. Furthermore, in case of $\bar{a}_f = 1, \bar{a}_f = 5$ and at low rotating speeds (relatively small bearing number Λ) something quite interesting occurs. The compliance of the overall bearing depends on the lubricant film which is rather softer than the so-assumed elastic foundation. This is demonstrated by the stiffness coefficient which tends to approach the same value, independent to the foil's compliance. In contrast, at high rotating speeds (or large bearing numbers) the compliance of the overall bearing depends on the elastic foundation. This is demonstrated by the stiffness coefficient which tends to be constant at high speeds independent to the foil's compliance. Finally, as it was expected in case of $\bar{a}_f = 0$ the four stiffness coefficients are solely a function of the lubricant film. These observations will be proved of significant importance, when the stability margins of the whole rotor-bearing model will be examined.

It is observed that the damping coefficient decreases as the compliance is increased due to the increased deflection of the foil. Furthermore, as stiffness coefficient clearly depicts, at low rotating speeds the compliance of the overall bearing depends on the lubricant film. This is the reason why all damping coefficients approach the same value for small bearing numbers independent of the compliance of the foundation. Finally, as one may observe damping coefficient decrease at higher rotating speeds due to the increased stiffness of the gas film which prevents energy dissipation.

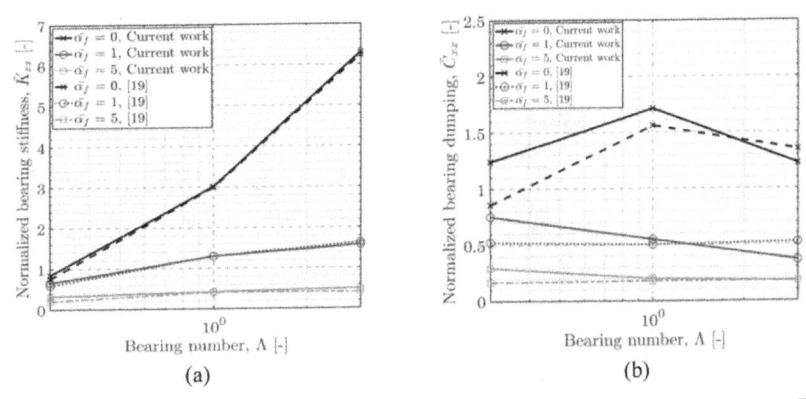

Fig. 3. Validation of linearized stiffness and damping coefficients for GFB: (a)$\overline{K}_{xx}$, (b)$\overline{C}_{xx}$

The most important validation test is the comparison of the dynamic characteristics of three different analytical models. The first analytical model is the reduced nonlinear model, which includes the Guyan reduction method and the Reynolds equation reduction method. The second model is the reduced linear one, according to which the bearing forces are evaluated via the four stiffness and four damping coefficients and the behavior of the rotor can be approximated by the static reduction method. Finally, the third analytical model is the full linear one, which just includes the linearized stiffness and damping coefficients of the bearings. In other words, this ensures that the Guyan reduction method, the Reynolds equation reduction method and the linearized stiffness and damping coefficients are adequately evaluated in this problem.

Table 2. Reference key properties of the rotor

Property	Value
Slenderness ratio, SR	20
Bearing span, L_s [m]	0.37
Young's modulus, E [GPa]	70
Rotor's mass, m_r [Kg]	2.51
Shaft's mass, m_s [Kg]	1.11
Natural frequency of the equivalent uniform, hinged Jeffcott rotor, ω_n [rad/s]	1000

The reference key properties of the rotor's analytical model are presented in Table 2.

In the beginning of the section, some constant parameters regarding the physical properties of the bearings and the number of mesh points in both the axial and the circumferential direction are given. To an extension, some extra properties regarding the static load and the bump foil compliance and loss factor are presented in Table 3.

Table 3. Gas foil bearing's extra properties for the validation process

Property	Value
Dimensionless static bearing load $\overline{W}_{st}$	0.5
Dimensionless bump foil compliance $\bar{a}_f$	1
bump foil loss factor η	0.1

In Fig. 4, the unbalance response at one gas-foil bearing node is depicted. For all the three cases depicted, single unbalance of grade G1 is considered. The unbalance grade is sufficiently low to keep the journal's orbit close to its equilibrium position. In the case of linear analytical models, the responses are evaluated via the linear harmonic Analysis and in the case of the nonlinear model the response is calculated via time integration considering low rotating acceleration. Sufficiently good agreement between the three models is observed until the threshold rotating speed of instability. Resonance frequencies are identical among the three models while resonance amplitudes appear very similar.

Linear harmonic analysis provides an estimation of the eigenvalues and the corresponding eigenvectors of the linear system. In the case of the nonlinear model, modal parameters are assessed via linearization around the respective speed depended equilibrium position. The comparison of the results is depicted in Fig. 4. Further, the validity of the derivation of the method for the linearized stiffness and damping coefficients is ensured via the results of Fig. 4.

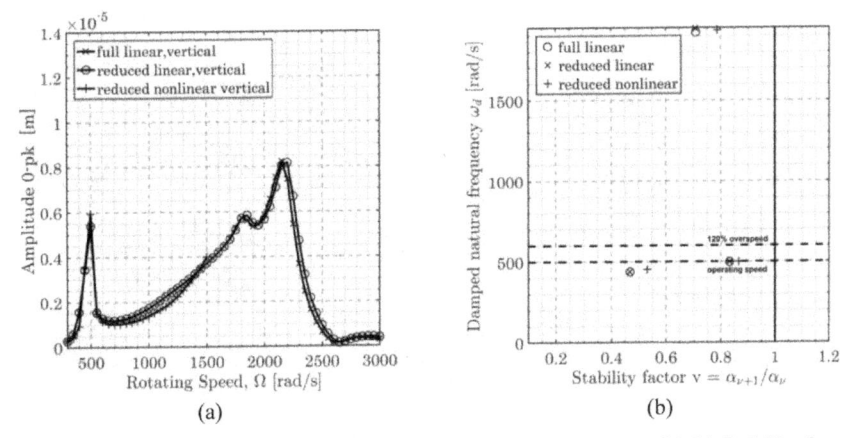

Fig. 4. (a) unbalance response of GFB#1 journal with single unbalance G1 (b) Stability factor v of the modes of the system operating at 500 rad/s.

Results regarding the implementation of parametric excitation are presented in continue. The capability of achieving the desirable amplitude ratios and excitation frequencies (for the stiffness and damping coefficients) under the effect of the predefined periodic deformation of the outer ring is investigated. Therefore, the main purpose is to determine the appropriate foil deformation in horizontal and vertical axis, dh and dv, see Appendix.

In Fig. 5, the variation of the stiffness coefficients for three different values of the bearing number and for three different cases of the excitation frequency is depicted. The bearing's properties are already presented in Table 1 and Table 3 while the horizontal maximum displacement of the outer ring is $dh = 0.01 c_r$. The normalization of the excitation frequency has been done using the rated rotating speed of the rotor $\Omega_r = 1000\ [rad/s]$.

It is expected that the mean value of the coefficients, calculated for low excitation frequency for a fixed bearing number, are equal to the corresponding coefficients of the bearing without excitation, due to the low deformation amplitude of the ring. One may observe this is not verified, due to the formulation of the vertical periodic load, see Appendix. Thus, the mean value of the ring's deformation is not equal to zero and the mean equilibrium position as well as the mean value of the stiffness and damping coefficients is affected. In addition, it is observed that all four stiffness coefficients reach their maximum value at $\Omega_{ext} t = \pi \ [rad]$, when the ring's deformation reaches its mean value with maximum negative radial velocity. This effect can be interpreted as follows. The stiffness coefficients depend on the journal's equilibrium position, which is mainly affected by the radial velocities of the deformable ring. Thus, when these velocities become negative, the equilibrium position is removed from the bearing's center and the corresponding stiffness coefficients are higher. Due to the same reason, the amplitude of all the stiffness coefficients increases as the excitation frequency increases too. It can be concluded that the direct couple terms are much more affected by the ring's deformation than the cross-couple ones.

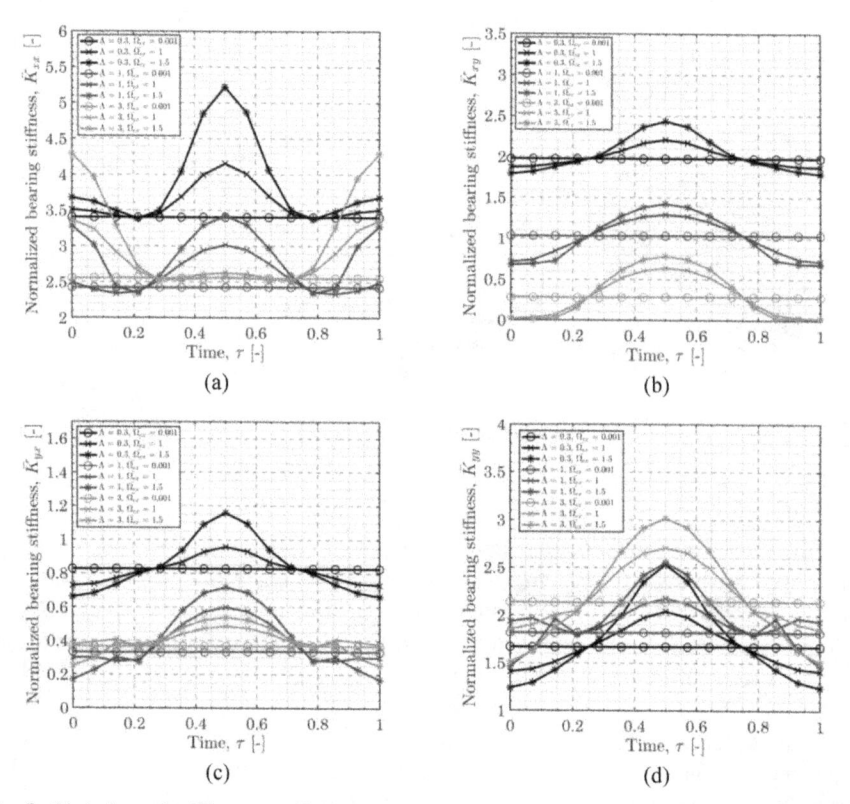

Fig. 5. Variation of stiffness coefficients under the effect of ring's periodic deformation with horizontal amplitude $dh = 0.01 \cdot c_r$. (a) $\overline{K}_{xx}$, (b) $\overline{K}_{xy}$, (c) $\overline{K}_{yx}$, and (d) $\overline{K}_{yy}$.

In Fig. 6 the corresponding damping coefficients are evaluated for the same operating conditions as above. The bearing's configuration as well as the ring's deformation remains the same. At low bearing numbers the compliance of the bearing depends on the lubricant film thus the main source of energy dissipation is the hydrodynamic gas film itself. At higher bearing numbers the compliance of the bearing depends on the structural compliance thus the most significant source of energy dissipation is the bump foil. The two different sources of energy dissipation are the reason why the damping coefficients do not systematically reach their minimum value at $\Omega_{ext} t = \pi$ [rad]. Further to that, the higher the excitation frequency is, the higher the variance of all damping coefficients is.

The stability maps for the linear and the nonlinear approach of the parametrically excited rotor-bearing model are presented in continue. At the case of the linear model, the bearing forces are calculated via the periodically changed stiffness and damping coefficients assuming no phase lug between them and assuming the same amplitude ratio for all of them, see Eq. (21). Even if these assumptions are rather an oversimplification, they provide an indication about the excitation frequency and amplitude which leads to parametric antiresonance. Then, according to the aforementioned desirable excitation frequency and amplitude, the nonlinear stability threshold is calculated, proving that parametric antiresonance is feasible in GFBs.

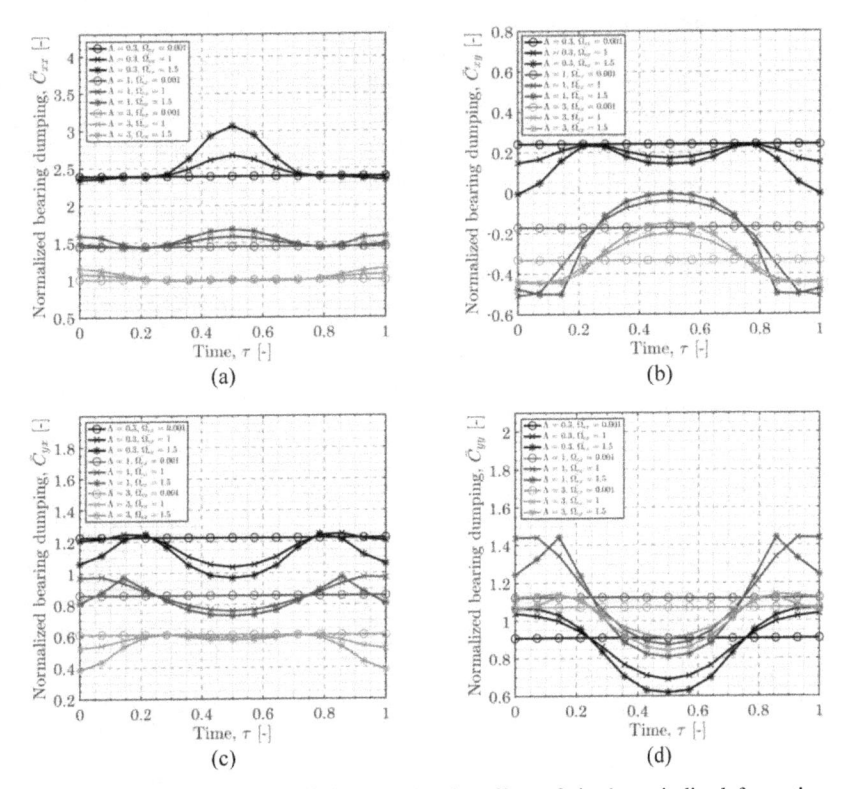

Fig. 6. Variation of damping coefficients under the effect of ring's periodic deformation with horizontal amplitude $dh = 0.01 \cdot c_r$. (a) $\overline{C}_{xx}$ (b) $\overline{C}_{xy}$, (c) $\overline{C}_{yx}$, and (d) $\overline{C}_{yy}$.

In Fig. 7, the maximum and minimum response at the respective periodic limit cycles at the horizontal degree of freedom, and at the first GFB is depicted for three ascending values of the amplitude ratio δ. The physical properties of the rotor and of the two identical bearings are already presented in Table 1 and Table 3. The range of both the excitation frequency and the rotating speed is selected in order to physically interpret the phenomenon of parametric antiresonance. Black dots denote stable periodic limit cycles and white dots denote unstable periodic limit cycles. All the periodic solutions have been evaluated via pseudo arc-length continuation method assuming a perfectly balanced rotor model (i.e. the only source of excitation is the periodically changing bearing properties). This continuation scheme is implemented accounting the rotating speed as bifurcation parameter. Pseudo arc-length continuation method incorporates orthogonal collocation. The latter determines the stability of the periodic solution as well as the type of the bifurcation (when occurring) through Floquet multipliers which are direct outcome from the collocation method (monodromy matrix).

Stable periodic limit cycles are found in both low and high rotating speeds regardless the excitation frequency. The corresponding stability analysis clarifies that the transition from stable to unstable limit cycles, in all excitation frequencies and rotating speeds, is occurring through secondary Hopf (Neimark-Sacker) bifurcation. At this case, the eigenvalues of the monodromy matrix which have magnitude greater than one present both real and imaginary non-zero part.

In Fig. 7, one may notice that as the excitation amplitude is increased, the Neimark-Sacker bifurcations approach each other and the stable periodic limit cycles become more. At frequencies close to antiresonance, the NS bifurcations collide and vanish. Close to antiresonance frequency, as the excitation amplitude is increased, the extent of the corresponding periodic limit cycles is decreased. Furthermore, it is important to note that the excitation frequency at antiresonance, can be approximately predicted by Eq. (25) where $\Omega_{cr,i}$ denotes the critical speeds of the system, approximately evaluated in Fig. 4. This notation is in full agreement with [16].

$$\Omega_f \simeq \Omega_{cr,2} - \Omega_{cr,1} \tag{25}$$

In Fig. 8, the maximum and minimum value of the periodic limit cycle response at the horizontal axis, and at the first GFB is depicted for two values of the excitation amplitude. The dimensionless foil compliance is $\bar{a}_f = 0.01$. The rotating system is stable when excitation frequency is inside a zone on the sides of antiresonance frequency.

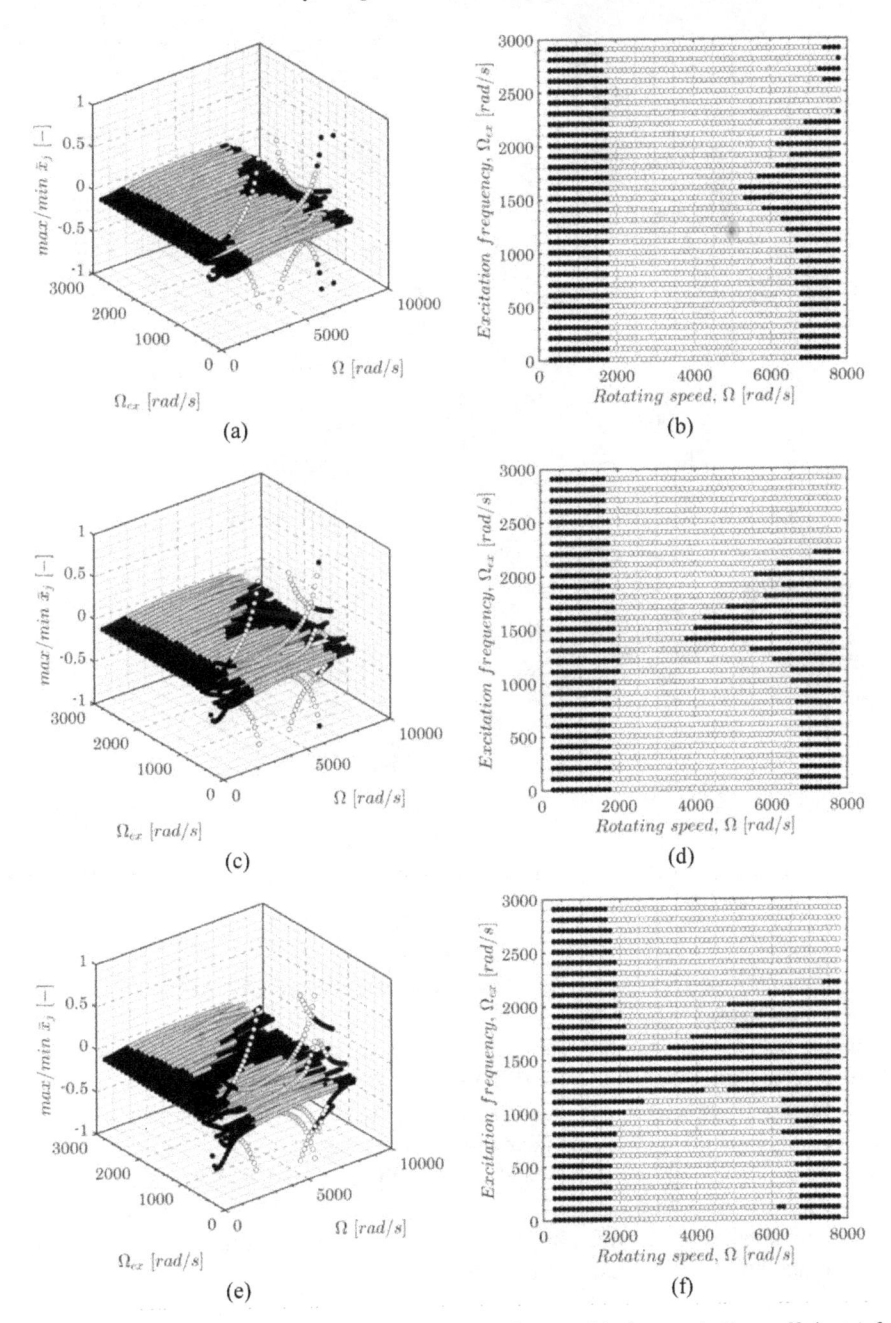

Fig. 7. Stability maps for the linear system (bearing forces with time-periodic coefficients) for (a) and (b): $\delta = 0.2$, (c) and (d) $\delta = 0.3$, (e) and (f): $\delta = 0.4$.

The Neimark-Sacker bifurcations collide and vanish. The extent of the periodic limit cycles evaluated in this zone of excitation frequencies is decreased. The current

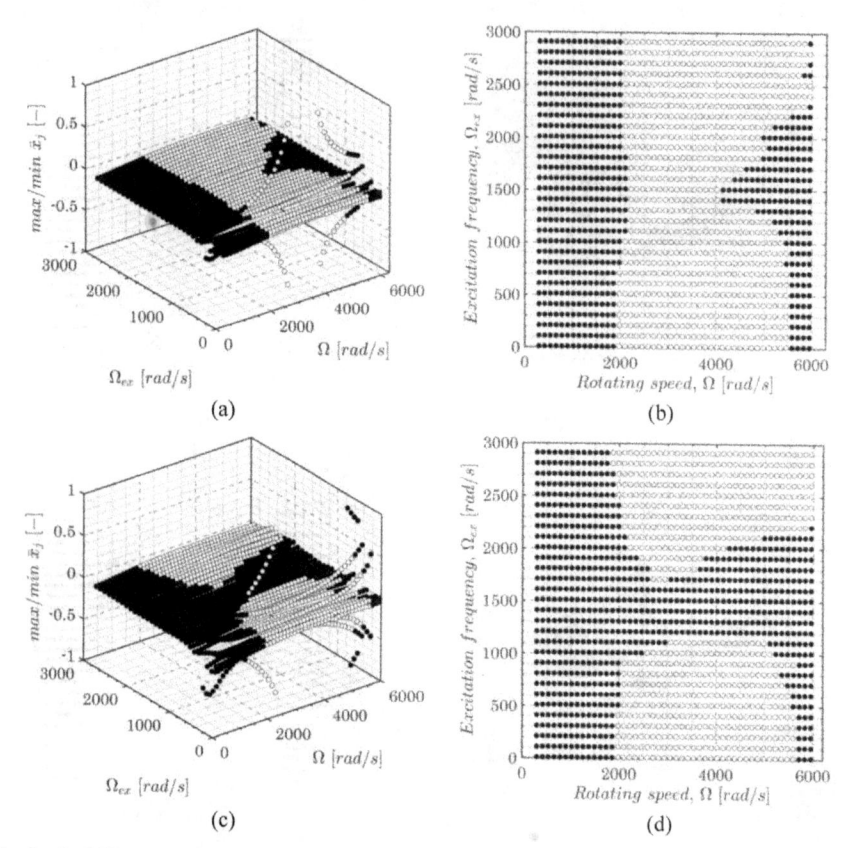

Fig. 8. Stability maps for the linear system (bearing forces with time-periodic coefficients) for (a) and (b): $\delta = 0.2$, and for (c) and (d): $\delta = 0.4$

configuration of the GFB is now utilized to evaluate the same periodic solutions with the nonlinear model, see Fig. 9.

In Fig. 9 the linear stability threshold is compared to the nonlinear one. Parametric antiresonance occurs also in the nonlinear model. The zone of excitation frequencies at which parametric antiresonance occurs strongly depends on the amplitude of excitation. Therefore, the significant discrepancies between the linear and the nonlinear stability threshold can be interpreted. Additionally, it should be noted that the method of evaluating the stiffness and damping coefficients for GFBs is not as accurate as possible, even for such high values of the foil compliance; this is the main reason that the comparison of linear and nonlinear models is presented in this work.

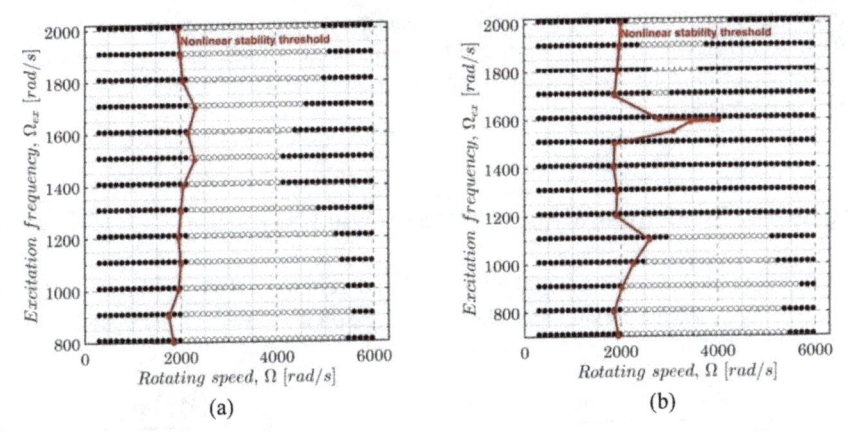

Fig. 9. Comparison of the stability threshold speed between the linear and the nonlinear system for (a)$\delta = 0.2$, and (b)$\delta = 0.4$

4 Conclusions

This work presents that parametric antiresonance is feasible in realistic slender rotors on active gas foil bearings. In this way, slender high-speed rotors can operate stable at higher speeds. Parametric excitation has been implemented by a vertical periodic load acting on the deformable ring of the gas foil bearings, whose elastoaerodynamic behavior is described by linear and a nonlinear approach. The stability and the extent of periodic limit cycles are evaluated via pseudo arc-length continuation method combined with an orthogonal collocation method. Based on the following conclusions this paper aims to raise further concerns on parametrically excited rotating systems.

According to a rather simplistic implementation of parametric excitation it is difficult to keep the same variation amplitude for all the stiffness and damping coefficients of GFBs. This fact does not enhance the parametric antiresonance phenomenon. The zone of the excitation frequencies in which parametric antiresonance occurs, depends strongly on the variation amplitude of the periodically changing stiffness and damping coefficients.

The investigation of the full bifurcation set at lower rotating speeds and higher excitation frequencies is recommended. In this paper, it was observed that period-doubling bifurcations occur at low rotating speeds independent from the excitation frequency. The evaluation of full bifurcation sets of such systems are ongoing work.

Parametric antiresonance can simplistically be interpreted as modal interaction. Consequently, it is quite interesting to evaluate the energy flow between the interacting modes (e.g. the 1st and 2nd bending mode) when parametric antiresonance occurs. This can be achieved by evaluating first the unbalance response of a parametrically excited unbalanced rotor. It should be noted, that in this case collocation method cannot be used as the majority of limit cycles will be quasi periodic due to the simultaneous unbalance (synchronous) and parametric excitation. Harmonic balance method is currently under development to be embedded in the corresponding continuation scheme.

Acknowledgment. The work in this paper is outcome of the ongoing research synergy between National Technical University of Athens and Karlsruhe Institute for Technology, entitled "Nonlinear Dynamics of Rotor Systems with Controllable Bearings", funded by the Alexander von Humboldt Foundation.

Appendix: Implementation of Parametric Excitation

The following analysis considers the physical and geometrical properties (inner /outer radius $R_{i,r}$, $R_{o,r}$, young modulus of elasticity, Poisson's ratio v_r) of the bearing ring as known and denotes:

$$\kappa_1 = 1 - \frac{\left(R_{o,r}^4 - R_{i,r}^4\right)}{2R_{i,r}^2\left(R_{o,r}^2 - R_{i,r}^2\right)} + \frac{1.33(1+2v_r)R_{o,r}}{\pi\left(R_{o,r}^2 - R_{i,r}^2\right)}, \quad \kappa_2 = 1 - \frac{\left(R_{o,r}^4 - R_{i,r}^4\right)}{2R_{i,r}^2\left(R_{o,r}^2 - R_{i,r}^2\right)} \quad (26)$$

Inspired by the analytically computed deformation of a ring under the effect of a periodic vertical load $F_0(1 + \sin(\overline{\Omega}_{ex}\tau))$, the author defines the horizontal and vertical deformation of the bearing ring as in Eq. (27) where F_0 denotes the amplitude of the periodic vertical load and I_r denotes the polar moment of inertia of the deformable ring.

$$q_r(\theta = 0, \pi, \tau) = dh = \frac{F_0}{2} \frac{R_{o,r}^3}{4.2 \cdot 10^{11} I_r} \left(\frac{2}{\pi}\kappa_2^2 - \kappa_2 + \frac{\kappa_1}{2}\right)(1 + \sin(\overline{\Omega}_{ex}\tau))$$

$$q_r(\theta = \pi/2, 3\pi/2, \tau) = dv = \frac{-F_0}{2} \frac{R_{o,r}^3}{4.2 \cdot 10^{11} I_r} \left(\frac{\pi}{4}\kappa_1 - \frac{2\kappa_2^2}{\pi}\right)(1 + \sin(\overline{\Omega}_{ex}\tau))$$

(27)

The corresponding derivatives with respect to the dimensionless time are given in Eq. (28).

$$d\dot{h} = \frac{F_0}{2} \frac{R_{o,r}^3}{4.2 \cdot 10^{11} I_r} \left(\frac{2}{\pi}\kappa_2^2 - \kappa_2 + \frac{\kappa_1}{2}\right)(\overline{\Omega}_{ex}\cos(\overline{\Omega}_{ex}\tau))$$

$$d\dot{v} = \frac{-F_0}{2} \frac{R_{o,r}^3}{4.2 \cdot 10^{11} I_r} \left(\frac{\pi}{4}\kappa_1 - \frac{2\kappa_2^2}{\pi}\right)(\overline{\Omega}_{ex}\cos(\overline{\Omega}_{ex}\tau))$$

(28)

The deformation of the outer ring and its rate of change are then evaluated in circumferential direction in Eq. (29), and in dimensionless form in Eq. (30).

$$q_r(\theta, \tau) = q_r = \sqrt{\left[(R_{i,r} + dh)\cos\theta\right]^2 + \left[(R_{i,r} + dv)\sin\theta\right]^2} - R_{i,r}$$

$$\dot{q}_r(\theta, \tau) = \dot{q}_r = \frac{\left[(R_{i,r} + dh)\cos\theta\right][d\dot{h}\cos\theta] + \left[(R_{i,r} + dv)\sin\theta\right][d\dot{v}\sin\theta]}{q_r + R_{i,r}}$$

(29)

$$\overline{q}_r = \frac{q_r}{c_r}, \quad \dot{\overline{q}}_r = \frac{\dot{q}_r}{c_r} \quad (30)$$

References

1. Bolotin, V.: The Dynamic Stability of Elastic Systems. Holden-Day (1964)
2. Seyranian, A.P., Mailybaev, A.A.: Multiparameter Stability Theory with Mechanical Applications. World Scientific Publishing Co. 13 (2003)
3. Schmidt,G.: Parametererregte Schwingungen (In German, Translated Title 'Parametrically Excited Oscillations'). Deutcher Verlag der Wissenschafte (1975)
4. Tondl, A.: On the interaction between self excited and parametric vibrations. Monographs and Memoranda, National Research Institute for Machine Design. 25 (1978)
5. Tondl, A.: To the problem of quenching self-excited vibrations. ACTA Technol. **43**, 109–116 (1998)
6. Dohnal, F., Ecker, H., Springer, H.: Enhanced damping of a cantilever beam by axial parametric excitation. Arch. Appl. Math. **78**, 935–947 (2008)
7. Breunung, T., Dohnal, F., Pfau, B: An approach to account for interfering parametric resonances and anti-resonances applied to examples from rotor dynamics. Springer Nature B.V (2019)
8. Dohnal, F., Chasalevris, A.: Improving stability and operation of turbine rotors using adjustable journal bearings. *Tribology International* (2016)
9. Boyaci, A., Hetzler, H., Seemann, W., Proppe, C., Wauer, J.: Analytical bifurcation analysis of a rotor supported by floating ring bearings. Nonlinear Dyn. **57**, 97–507 (2009)
10. Boyaci, A., Lu, D., Schweizer, B.: Stability and bifurcation phenomena of Laval/Jeffcott rotors in semi-floating ring bearings, Nonlinear Dyn. **79**, 1535–1561(2015)
11. Van Breemen, F. C.: Stability analysis of a Laval rotor on hydrodynamic bearings by numerical continuation: Investigating the influence of rotor flexibility, rotor damping and external oil pressure on the rotor dynamic behavior, M.Sc. thesis, Delft University of Technology (2016)
12. Rubel, J.: Vibrations in nonlinear Rotordynamics Dissertation, PhD thesis, Ruprecht-Karls-Universität Heidelberg (2009)
13. Amamou, A., Chouchane, M.: Bifurcation of limit cycles in fluid film bearings. Int. J. Non-Linear Mech. **46**, 1258–1264 (2011)
14. Sghir, R., Chouchane, M.: Prediction of the nonlinear hysteresis loop for fluid-film bearings by numerical continuation. Proc. IMechE Part C: J Mech. Eng. Sci. **229**(4), 651–662 (2015)
15. Sghir, R., Chouchane, M.: Nonlinear stability analysis of a flexible rotor-bearing system by numerical continuation. J. Vib. Control **22**(13), 3079–3089 (2016)
16. Becker, K.: DynamischesVerhaltenhydrodynamischgelagerterRotorenunterberücksichtigungveranderlicherLagergeometrienm Ph.D. Thesis, Karlsruhe Institute of Technology, Germany (2019)
17. Leister, T.: Dynamics of Rotors on Refrigerant Lubricated Gas Foil Bearings, Ph.D. Thesis, Karlsruhe Institute of Technology, Germany (2021)
18. Dohnal, F.: Optimal dynamic stabilisation of a linear system by periodic stiffness excitation. J. Sound Vib. **320**, 777–792 (2009)
19. Peng, J.P., Carpino, M.: Calculation of stiffness and damping coefficients for elastically supported gas foil bearings. J. Tribol. **115**, 20–27 (1993)
20. Heshmat, H., Walowit, J.A., Pinkus, O.: Analysis of gas-lubricated foil journal bearings. J. Lubr. Technol. **105**, 647–655 (1983)
21. Baum, C., Hetzler, H., Schroders, S., Leister, T., Seemann, W.: A computationally efficient nonlinear foil air bearing model for fully coupled transient rotor dynamic investigations. Tribol. Int. **153**, 10434(2020)
22. Allgower, E.L., Georg, K.: Introduction to Numerical Continuation Methods. Society for Industrial and Applied Mathematics, Society for Industrial and Applied Mathematics (2003)

23. Meijer, H., Dercole, F., Olderman, B.: Numerical bifurcation analysis. In: R. A. Meyers (ed.) Encyclopedia of Complexity and Systems Science, Springer New York, pp. (6329–6352). https://doi.org/10.1007/978-0-387-30440-3_373
24. Kuznetsov, Y. A.: Elements of Applied Bifurcation Theory 2nd ed., Applied Mathematical Sciences, Springer, New York (1998). https://doi.org/10.1007/978-1-4757-3978-7
25. Nayfeh, A.H., Balachandran, B.: Applied Nonlinear Dynamics. Wiley series in nonlinear science. Wiley, Hoboken (1995)
26. Doedel, E.J., Keller, H.B., Kernevez, J. P.: Numerical analysis and control of bifurcation problems (II) Bifurcation in infinite dimensions. Int. J. Bifurcat. Chaos **1**(3), 745–772(1991)
27. Doedel, E.J.: Lecture notes on numerical analysis of nonlinear equations. In: Krauskopf, B., Osinga, H.M., Galán-Vioque, J. (eds.) Numerical Continuation Methods for Dynamical Systems. UCS, pp. 1–49. Springer, Dordrecht (2007). https://doi.org/10.1007/978-1-4020-6356-5_1
28. Bhore, S.P., Darpe, A.K.: Nonlinear dynamics of flexible rotor supported on the gas foil journal bearings. J. of Sound and Vib. **332**, 5135–5150 (2013)

Author Index

B
Bitner, Alexander   65

C
Chantoumakos, Georgios   82
Chasalevris, Athanasios   134, 188
Chatterton, Steven   111, 150

D
da Silva, Heitor A. P.   44
Dassi, Ludovico   150
Dimou, Emmanouil   188
Dohnal, Fadi   188

F
Frigaard, Mads Nowak   1

G
Gavalas, Ioannis   134
Gheller, Edoardo   111

J
Jensen, Janus Walentin   27

K
Kalligeros, Christos   82
Kazakov, Yuri   12, 55

K (cont.)
Kornaev, Alexey   12
Koufopanos, Vasileios-Menelaos   162

L
Li, Shengbo   12

N
Nicoletti, Rodrigo   44
Nikolakopoulos, Pantelis G.   162

P
Papadopoulos, Anastasios   134
Parenti, Paolo   150
Pennacchi, Paolo   111, 150
Polyakov, Roman   12
Proppe, Carsten   65

S
Santos, Ilmar Ferreira   1, 27
Savin, Leonid   12, 55
Shutin, Denis   12, 55
Spiridakos, Panagiotis   82
Spitas, Vasilios   82

T
Tsolakis, Efstratios   82

V
Vania, Andrea   111, 150

© The Editor(s) (if applicable) and The Author(s), under exclusive license to Springer Nature Switzerland AG 2023
A. Chasalevris and C. Proppe (Eds.): ABROM 2022, LNME, p. 209, 2023.
https://doi.org/10.1007/978-3-031-32394-2